SUSTAINABLE DEVELOPMENT OF WATER RESOURCES IN NORTHEAST INDIA

Sustainable Development of Water Resources in Northeast India

Edited by

J.B. Bhattacharjee
Abhik Gupta
Niranjan Roy

AKANSHA PUBLISHING HOUSE
NEW DELHI • GUWAHATI • VISAKHAPATNAM

AKANSHA PUBLISHING HOUSE
4649-B/21, Ansari Road
Daryaganj, New Delhi-110002
Phones: 23268193, 09435116718
Email: ektabooks@yahoo.com
visit us: www.akanshapublishinghouse.com

Sustainable Development of Water Resources in Northeast India

First Published 2017

ISBN 978-81-8370-272-0

PRINTED IN INDIA

Published by Akansha Publishing House, New Delhi-110002 and printed at Harmain Offset Printers, Delhi.

Contents

Preface

Water covers about 70 per cent of the earth, although only 3 per cent of this is freshwater and only 13 per cent of this 3 percent is accessible, the rest being locked up in ice sheets and rocks. This 13 per cent has also a highly uneven distribution over the surface of the earth and at different times of the year. The irony of the situation is that even places like Cherrapunjee or Mawsynram, which experience the highest rainfall in the world, also suffer from acute water shortages in winter, as most of the rainfall occurs during the monsoon and pre-monsoon, leaving some months of the year water deficient. Water, therefore, needs efficient management for its conservation, storage and distribution. An additional dimension that has been added in the recent years is the quality of water. Globally, about two million tonnes of human waste are deposited into the various water courses. These wastes not only pollute the surface water, but also percolate down to contaminate the ground water. Microbial contamination and nitrate pollution of groundwater are important examples of anthropogenic (i.e. human-induced) problems of groundwater management. Sometimes, naturally present toxic substances like arsenic and fluoride also pose serious health hazards to groundwater users and impose severe restrictions on its use. Besides these chemically induced problems, bacterial and viral contamination of water remain the greatest hazard with an estimated annual incidence of 4.6 billion diarrhoea episodes that cause 2.2 million deaths every year in the world. Over 1 billion people in the world do not have access to safe drinking water and about 33 per cent of the world population does not have proper sanitation facilities. Diarrhoea is still the second greatest cause of mortality of the children. A UNICEF report reveals that diarrhoea kills around 4900 children every day, of which, as revealed by a 2007 report of the Ministry of Health, Government of India, around 1000 deaths are from India alone. In other words, more than 41 children die in India every minute. This is a matter of grave concern and a pointer to the long way our country has to travel to reach one of the millennium development goals of substantially reducing child mortality.

Northeast India now comprises eight states after the recent inclusion of Sikkim, the earlier 'seven sisters' being Arunachal Pradesh, Assam, Manipur, Meghalaya, Mizoram, Nagaland and Tripura. All these eight states lay within the Brahmaputra and Meghna (Barak) Drainage Sub-basins of the Ganga-Brahmaputra-Meghna Basin. Besides draining Arunachal Pradesh and parts of Assam, Nagaland and Meghalaya in India, the Brahmaputra and its numerous tributaries also drain the whole of Bhutan and parts of China (Tibet) in the upstream and parts of Bangladesh in the downstream. The Barak, on the other hand, drains parts of Nagaland, Manipur, Assam, Mizoram and Meghalaya in India and the Surma-Meghna tracts in Bangladesh. The Brahmaputra Sub-basin has an average annual flow of 629.05 cu km per year, and Barak an average annual flow of 48.36 cu km, making the total flow 677.41 cu km, which is even more than the 525.02 cu km of the Ganga Sub-basin, and is the highest for any river basin in India. However, because of topographical and other constraints, the utilizable flow of Brahmaputra-Barak is a mere 24 cu km when compared with the 250 cu km of the Ganga, and less than a number of other rivers like Krishna, Godavari, Mahanadi and Indus. After adding an utilizable groundwater potential of 22.15 cu km, the total utilizable flow comes to 46.15 cu km. Moreover, 66-85 per cent of the total rainfall in this region occurs during the four monsoon months of June-September, leading to frequent flooding in both Brahmaputra and Barak valleys accompanied by the destruction of crops, property and life. On the other hand, there is considerable water scarcity during December-March every year, and the state like Mizoram has to officially adopt rainwater harvesting in the monsoon to tide over the lean season shortfall. Several hill cities, like Shillong, Aizawl, Kohima and others, suffer to a varying extent from drinking water shortages. It is also a known fact that very little development and utilization of groundwater resources have taken place in this region, barring perhaps Tripura which has so far utilized about 33 per cent of its utilizable groundwater potential. In Assam and Meghalaya, this figure is 8.75 and 3.97 per cent respectively, while it is negligible in the other states. However, groundwater exploitation in many areas faces the risk of arsenic and fluoride contamination, besides widespread high iron levels.

The state of Assam, the largest in the Northeast, is comprised of three physical divisions, viz., the Brahmaputra Valley, the Barak Valley and the Hill Range. The two valleys derived their names from the respective main rivers, the Brahmaputra and the Barak, flowing in east to west direction. The hill division formed by the districts of Karbi-Anglong and the North-Cachar (Dima Hasao) hills acts as the physical divide between the two valleys. The Barak Valley, i.e. the area on focus in this collection, is situated between Latitude 24^0 8¢ and 25^0 8¢ North and Longitude 92^0 15¢ and 93^0 15¢ East, covering an area of 6922 sq. km. It constitutes 8.9 percent of the geographical area of Assam but contains 11.22 percent of its population (2001 census) and shares its border with North-Cachar hills district and the state of Meghalaya in the north; state of Manipur in the east; state of Mizoram in the south; and the state of Tripura and the Sylhet district of Bangladesh in the west. Administratively, it comprises of three districts, namely, Cachar, Karimganj and Hailakandi, with geographical areas of 3786, 1809 and 1327 sq. km, respectively. Silchar, Karimganj and Hailakandi are the district headquarters. Cachar district is comprised of two sub-divisions, namely, Silchar and Lakhipur, while other two districts are without any sub-division. Physiographically, the Barak Valley can be divided into four major landforms; viz., undulating plains, broad meander plains, flood plains and low-lying areas. The undulating plain areas scattered across the region are mixed with low hills and meander plains; broad meander plain occurs mainly on the north bank of the Barak river in large patches mixed with low hills and piedmonts; flood plains area chiefly covers the banks of the Barak river; while, the low-lying areas consist of natural depressions and waterlogged terrains. These landforms are scattered in all the three districts and occur mainly on the south bank of the Barak river which passes through all the districts in the valley.

River Barak, which originates a little to the west of Maothana on the northern boundary of Manipur, takes a south-west turn near Tipaimukh in Manipur and flows northwards, forming for a considerable distance the boundary line between Assam (Cachar district) and Manipur, where it is joined by river Jiri, and then turns to the west near Jiribum. Traversing through Cachar and Hailakandi districts, the Barak reaches Badarpur town in Karimganj district, forms the boundary between Karimganj district and Bangladesh from Badarpur to Haritikar, where the river divides into two branches, namely, Surma and Kushiara (Kosiyara). While Surma immediately enters Bangladesh, Kushiara flows further by Karimganj town and then enters Bangladesh. Within the Indian portion of the valley, Barak receives numerous tributaries from the neighbouring hills like Chiri, Madhura, Jatinga, Dalu, Harang, etc. on its north bank and Sonai, Ghagra, Dhaleswari, Katakhal, Longai, Singla and Kakra on its south bank. Most of these rivers dry up in winters and their drainage capacities are drastically reduced due to deposits of silts, but in the rainy season all of them rise up in high spate causing devastating floods almost every year. The widespread water logging and several rounds of floods is a consequence of heavy rainfall in the region during the summer months. The floods cause enormous loss of life, crop and property. Most floodplains in the valley are in the form of low-lying areas with numerous small hillocks strewn within them. A unique feature of the floodplains is that when they are inundated during the monsoon months of June-September, they are treated as common property fishing grounds by the community but after the water recedes, private property rights are re-established in the agricultural fields. In spite of having an annual rainfall of 2500-3000 mm, the floodplains in the Barak Valley experience water scarcity during the dry months of December-March, as irrigation facilities are almost non-existent in the area. The heavy pre-monsoon rains in April-May also often lead to flooding that destroys the crop. In this paradoxical situation the people are indeed caught in a vicious cycle of either too much or too little water. The rampant landlessness among the peasants, a general decrease in the size of landholding and the overexploitation and depletion of fish stock are among other human problems of the area which are attributable to the burgeoning population pressure. The worst fact is that despite being richly endowed with water resources, the Barak Valley region has not so far been able to utilize this precious natural resource for economic development; instead, this resource has been causing a perennial problem for the entire valley in the form of annual occurrence of devastating floods. Therefore, the sustainable water management appears to be pivotal for ameliorating the socio-economic conditions of the people in this ecologically and economically important landscape.

Institute of Northeast India Studies, which is committed to its mission to conduct studies and research in areas relevant to socio-economic and cultural advancement of the people in the north-eastern region of India, identified environmental issues concerning the region as one of the research priority areas. To sensitize the people and the government about their respective role in view of the global warming and impending threat of climate change is within its agenda. A brainstorming session at the Assam University, Silchar, in August 2008, on 'Development Strategies for Barak Valley (Assam)' convinced us that sustainable water management should indeed be the starting points for all development strategies for the region. A symposium on the sustainable development of water resources was accordingly planned with the broad objectives (i) to make a realistic assessment of the water resource potential; (ii) to analyze the extents of utilization and wastage of water resources; and (iii) to formulate a sustainable policy package for the development of water resources in the region.

Sustainable Development of Water Resources in Northeast India thus emerges from a symposium on 'Sustainable Development of Water Resources in Barak Valley' organized by the Silchar Centre of the Institute of Northeast India Studies (INEIS) at Assam University, Silchar, on 24th March 2010. Dr. Abhik Gupta, Professor of Ecology & Environmental Science, Assam University, and Dr Niranjan Roy, Director, Internal Quality Assurance Cell (IQAC), organized the symposium as Convener and Joint Convener respectively. Professor Tapodhir Bhattacharjee, Vice Chancellor, Assam University, chaired the inaugural session. Professor G.D. Sharma, Pro-Vice Chancellor, Assam University, and Professor B.P. Misra, formerly Director, Centre for Himalayan Studies, North Bengal University, were the Guests of Honour. Professor Shib Ranjan Misra, Professor of Economics, Visva-

Bharati University, delivered the Keynote Address. The academic sessions were chaired by Professor G.D. Sharma and Professor Shib Ranjan Misra. In keeping with the tradition of INEIS for all academic programs, the open house interaction in the evening, presided over by the INEIS Chairman, was a lively discourse on the magnitude of the problem and the various alternatives for sustainable development of water resources proposed in the academic sessions. About eighty scholars, media persons and social activists representing NGOs, including Life Members of INEIS from the Silchar area, actively participated in the symposium till the end of the interactive open house session very late in the evening. The open house observed that, although the emphasis of the symposium was initially on the Barak Valley area and the draft papers were prepared by the members of faculty and research scholars in Assam University on the basis of localized case studies, the north-eastern regional dimension was very strong in most of the presentations and that for the purpose of sustainable development it is but natural that the academia requires an integrated approach for studying the Northeast holistically. The proceedings volume should, therefore, address the entire region, though there may remain some gap areas to be filled in by similar discourses in future. While the recommendations of the participants were loud and clear about the need to sensitize the masses on consequences of global warming and threats of climate change and to involve the local people in conservation of water bodies and wetlands in the region, there was also a near unanimity in proposing that indigenous knowledge and technology can greatly help the sustainable development of water resources in the Northeast, that multiple minor and medium river-canal projects can irrigate agricultural fields and booast power generation, that hydro-projects need not be looked upon as the only means for power generation, that the dam projects should not be undertaken without studying the environmental and socio-economic implications and obtaining informed consent of the stakeholders, and that small and medium scale dams should be the better options in this biodiversity hotspot. The presenters were asked to re-look their drafts in light of the comments and suggestions of the participants and to submit the papers for publication. Indeed, they cooperated very well by maintaining the given dateline in handing over the final drafts. The book thus contains the revised papers of the symposium. It begins with the illuminating keynote address of Professor Shib Ranjan Misra which had set the tune of the symposium.

I am indeed thankful to Professor Tapodhir Bhattacharjee, Vice-Chancellor, Assam University and Chairman, Silchar Centre of INEIS, and his colleagues in the University and the Centre, for organizing the symposium for socio-economic and cultural benefits of the people of the region that the Assam University is committed to serve by its own mandate. The Internal Quality Assurance Cell (IQAC), Assam University, which is striving hard towards ensuring quality research in the University and Community-University participation in socially relevant research initiatives, offered logistic support for success of the symposium. In fact, the presentations in the symposium were by the faculty members and research scholars of the University based on their own research on water resource related development issues of the region which currently seems to be at the centre of various strategies for sustainable socio-economic and cultural development in the Northeast. Dr. Abhik Gupta, Professor of Ecology & Environmental Science, and Dr. Niranjan Roy, Professor of Economics & Director, IQAC, Assam University, associated themselves with the Chairman, INEIS, in jointly editing the book. I am also thankful to all participants of the symposium and to the authors whose contributions have made this publication possible. Finally, I reiterate my thanks to Shri M. P. Misra of the Akansha Publishing House, New Delhi, who has been so consistent in his support to the initiatives of INEIS in promoting Northeast India Studies. It is hoped that the book will generate interest in studying, conserving and developing the water resources for sustainable development of the Northeast.

Institute of Northeast India Studies
Kolkata - 700129
January 15, 2011

—J.B. Bhattacharjee

Our Contributors

1. Dr. Shib Ranjan Misra, Professor, Department of Economics, Visva-Bharati University, Santiniketan – 731235.
2. Dr. Abhik Gupta, Professor, Department of Ecology & Environmental Science, Assam University, Silchar – 788011.
3. Dr. G.D. Sharma, Professor, Department of Life Science & Pro-Vice-Chancellor, Assam University, Silchar- 788011.
4. Dr. (Ms) Tapati Das, Assistant Professor, Department of Ecology & Environmental Science, Assam University, Silchar – 788011.
5. Dr. Niranjan Roy, Associate Professor, Department of Economics & Director, IQAC, Assam University, Silchar – 788011.
6. Dr. (Mrs.) Suparna Roy, Assistant Professor, Department of History, Assam University, Silchar – 788011.
7. Ms. Kankana Das, Research Scholar, Department of Ecology & Environmental Science, Assam University, Silchar – 788011.
8. Dr. (Mrs.) Susmita Gupta, Assistant Professor, Department of Ecology & Environmental Science, Assam University, Silchar – 788011.
9. Ms. Hafsa Sultana Laskar, Research Scholar, Department of Ecology & Environmental Science, Assam University, Silchar – 788011.
10. Dr. (Mrs) Mithra Dey, Reader, Department of Ecology & Environmental Science, Assam University, Silchar – 788011.
11. Dr. Nikhil Bhusan Dey, Professor, Department of Commerce, Assam University, Silchar – 788011.
12. Dr. Pranay Jyoti Goswami, Associate Professor, Department of Commerce, Assam University, Silchar – 788011.

13. Mr. Bimal Sinha, Research Scholar, Department of Business Administration, Assam University, Silchar – 788011.

14. Ms. Anuradha Singha, Research Scholar, Department of Economics, Assam University, Silchar – 788011.

15. Mr. Ananta Pegu, Research Scholar, Department of Economics, Assam University, Silchar – 788011.

16. Dr. Parthankar Choudhury, Assistant Professor, Department of Ecology & Environmental Science, Assam University, Silchar – 788011.

17. Mr. Pijush Kanti Das, Research Scholar, Department of Bengali, Assam University, Silchar – 788011.

18. Dr. Projit Kumar Palit, Associate Professor, Department of History, Assam University, Silchar – 788011.

19. Mr. Bhaskar Singha, Research Scholar, Department of Economics, Assam University, Silchar – 788011.

1

Sustainable Water Management in Northeast India with Special Reference to Barak Valley: Some Emerging Issues and Concerns

Shib Ranjan Misra

[**Abstract:** Northeast India has a unique topography and it is endowed with abundant water resources. In the Barak valley of Assam, the river Barak and its tributaries can provide abundant scope for sustainable agriculture and fisheries development, providing a stable economic base for livelihood. But due to poor irrigation facilities and other weak management practices, the region suffers precisely where rainfall is insufficient and irregular, and excess and sudden heavy showers lead to water logging and flash floods. Rainwater harvesting, proper management of existing water resources, watershed development and community participation will help attain sustainable utilization of water for agriculture and uplift the socio-economic conditions of the people. The key words are Sustainable development, Tragedy of the commons, Watershed management, Big dams, Climate change, People's participation.]

Water is the essence of life. The water resources in the world are fast depleting and consequently, increasing water scarcity is a serious threat to life. In many countries including India, water availability is likely to approach the critical stress level soon. It is to be noted that India has an enviable record of engineering accomplishments in the field of irrigation and those accomplishments deserve much of the credit for the nation's impressive growth in foodgrain production since its Independence. The fact that irrigation is essentially a post-Independence phenomenon is quite clearly recognized from the investment pattern in the successive Five-year Plans. A lion's share (9.78 per cent) of such investment has been on major and medium irrigation projects. But potentials created under different schemes have not been fully utilized.

Water is the limiting factor for crop production, and water management is the key to sustainable agricultural development. Though India gets an average of 75 cm annual rainfall, its distribution in space and time makes Indian agriculture a gamble in monsoon. It is never too late to recognize that land and water form the basis of any sustainable system of agriculture and their improvement should lay the foundation of productive, economically viable and socially acceptable agricultural system. The future of agriculture lies in recognizing the basic truth that most of the food, fuel, fibre and forage will be produced from land and water. The water management technologies are required for continuous enhancement of productivity and sustainability of water.

An important consideration that should have to be taken care of while formulating the strategy for water resource management in a country like India has been the recognition of the need for production of food, fodder, fibre, fuel, industrialization and urbanization. All these put severe competing claims on land use as well as water use. Further, it will be required to consider the water demand for livestock population. Unless the water use efficiency improves to a reasonable extent, the lack of fresh water is likely to be a major physical constraint for future food production and development in the country. Again, the frequent occurrence of drought and famines in the country has heightened the need to utilize the available water fully and perfectly, particularly for the rainfed farming which is fraught with multiple risks. Farmers operating in arid and semi-arid areas under rainfed conditions face, more than most others, a multitude of perceived and oft-experienced risks of varying severity emanating from the uncertainties inherent in their natural, economic and socio-economic environments. It is in this context that irrigation becomes important to provide insurance against possible risks in farming. Several forms of irrigation interventions for drought have been widely heralded.

Again, there is a view emerging where it is held that irrigation, surface or ground water, has proved a costly experiment. If one accepts the view, then the immediate question: what are the factors that lead to the emergence of such a view? Till recently, irrigation has been considered synonymous with the construction of dams and canal network. Many irrigation projects have been designed and managed almost exclusively from an engineering point of view (Thavaraj: 1982). The socio-economic and environmental dimensions do not seem to have been integrated with technical design. So, it is sometimes argued that it is not correct to characterize the irrigation projects as purely hydraulic systems to be run according to engineering principles (Misra : 2006). All participants – like farmers, irrigation officials, and politicians - are expected to play their role to maximize their private interests to the cost of social goods (Caruthers : 1987). It is, in essence, an integral part of socio-economic system.

Essentially, the policies of irrigation water distribution and use appear to have not taken into account the responses and understanding of the cultivators. The need for understanding the nexus between the technology *per se* and its operationalisation has a special significance in a democratic system. Caste, class and factional ties and antagonisms have become relevant in the delivery of any goods and services to the community (Srinivas : 1984). The mismatch, if any, between theoretically assumed policies and goals and actual results in a project leads to a chain of adverse socio-economic and environmental problems. Quintessentially, without understanding the interconnection and interdependence of various ecological entities like land, water, flora, fauna and human along with human syndrome, no worthwhile water management system is possible. In fact, the over-domination of the engineering-centred capital-intensive project-oriented approach to water planning is the bane of the crisis. The societal vision is totally missing in the technical and technological solutions that are generally perpetrated. There are certain issues that should be noted. These are broadly concerning the kind of irrigation works that should be promoted, the relative importance that should be given to quantitative expansion, the means of achieving qualitative improvement, the question of equity in the distribution of benefits from irrigation, environmental aspects, and the societal vision

There is a big question of poor economic returns from major irrigation projects and the negative effects on environments resulting in soil degradation, water-logging and salinization in many areas. In India, the percentage of area damaged by poor water management is very high. The worst problems are the declining water table, rapidly depleting ground water resources. Grave concerns are being expressed about ill effects of poor management of irrigation on ecology of the command areas of many irrigation projects. In all fairness, it is pertinent to see that the irrigation systems should be managed rather than administered. The operation and maintenance of the system should be given priority.

The state of Assam, the largest state of the north-eastern region of India, is highly endowed with large number of water resources. She covers an area of 78,438 sq km and consists of two valleys i.e. the Brahmaputra valley and the Barak valley. The two valleys derived their names from the respective main rivers, the Brahmaputra and the Barak, flowing through the valleys in an east to west direction. River Barak of Assam – the second largest river in the Northeast Region of India - sustains millions of people. It is the perennial source of water to the people of Barak valley, providing various ecological and economic services. Barak is also the main river which maintains the balance of peripheral water bodies including lentic eco-systems and all plans of sustainable water usage that have been initiated. The rivers like the Barak cause major problems during the monsoon period of every year in the shape of flood, bank erosion and drain congestion. Most of the rivers in the Barak valley dry up during winter and due to deposits of silt, their drainage capacities are drastically reduced. The rivers remain navigable only in the rainy season. As such, in summer all these rivers rise up in high spate causing floods almost every year. In fact, the successive waves of devastating floods in almost every year have virtually destroyed the economy of the valley. People living in the low lying areas are forced to leave their homes. Heavy downpour submerges the villages throwing normal life out of gear. Danger looms large due to overflowing. Thousands of people are marooned as heavy rain lashes the area. Rivers, like the Barak, Khusiara and Langai are recorded flowing in most of the years well above the danger mark. Every year the floods leave a trail of destruction, washing away villages, submerging paddy fields, drowning livestock, besides causing loss of human life and property cost in several billions. For Barak valley, the mighty Barak river, though a life-giving river, has become more synonymous with devastation than with prosperity. In fact, the successive waves of devastating floods in almost every year have virtually destroyed the agriculture-based economy of the region. To remove the backwardness of the region, tackling the flood menace is a must. Otherwise, all efforts to bring this region at par with other mainland states will come to a naught.

The main factors causing extensive floods are the adverse physiography of the region, heavy rainfall, excessive sedimentation, frequent occurrence of the earthquakes, hill land sliding, reduction of forest area and encroachment of the rivers in many areas. Usually, Assam experiences incessant rainfalls during the monsoon season, which normally commence from the month of July and remain till mid-October. Apart from this, occurrence of floods in Assam has direct correlation with rainfall in the catchments areas of neighouring states of Arunachal Pradesh, Nagaland, Meghalaya, Mizoram, Manipur, etc. Most floodplains in the Barak valley are in the form of low lying areas with numerous hillocks. They get inundated during the monsoon months of June-September. A unique feature of the floodplains is that when they are inundated during monsoon, they are treated as common property fishing grounds by the community; while after the water recedes, private rights are re-established in the agricultural plots. Thus, it appears that the situation is quite paradoxical, because in spite of having an annual rainfall of 2,500-3,000 mm, the floodplains experience water scarcity during dry months of December-March. Again, pre-monsoon rains in April-May often lead to flooding that destroys the crop. Thus the people are caught in a vicious circle of either too much or too little water.

Thus despite being richly endowed with water resources, the Barak valley region could not utilize this resource for economic development. This resource has created a perennial problem for the entire valley in the form of occurrence of floods. Against this backdrop, there is no denying the fact that sustainable water management appears to be pivotal for improving the socio-economic conditions of the people of this ecologically and economically important landscape of Assam. In fact, the apathy for harmonious use of land and water in accordance with potentiality and capability of soils has given rise to serious problems. They require appropriate management.

Due to the increasing population multiplied by their desire to attain higher economic growth, the Barak river is increasingly polluted through establishment of industries nearby the river. The over-flooding of Barak annually is further aggravated by the deforestation by human activities in the form of soil erosion, sand casting, sedimentation, etc. The problem is primarily of anthropogenic nature. It is the human who in the urge for attaining higher growth and development exploit the very basis of the development. On the other hand, due to the global warming, changes have been brought into the hydrological realm of the rivers and the weather and the climate. Water resources in the Barak valley are facing the similar problems due to the natural and human made forces. The solution lies in realization of the fact that the water is the natural capital which needs to be conserved for the development of the region.

In Assam, more particularly in the Barak valley region, the flood and erosion cause enormous damages to the crops, livestock, land property and bring untold miseries to the people at large. The floods cause large inundation and spread devastation to standing crops, homestead, life and property, disruption of road and rail communication, public utilities, water supply installations, irrigation and flood control structures, etc. The National Flood Commission had estimated the area vulnerable to floods in Assam at 31.60 lakh hectares against 335.16 lakh hectares for the whole of India. Assam thus accounts for 9.4 per cent of total flood prone area of the country. As a result, the state has not been able to achieve the desired progress and prosperity in spite of having vast natural resources. Floods in three to four waves is an annual feature, which, very often, wipes out major chunks of the fruits of the people's labour. The Table below shows the extent of damages caused by floods in the state of Assam over the years 2000-2004.

Table: Damages caused by Floods in Assam

Items	*Unit*	*2000*	*2001*	*2002*	*2003*	*2004*
Area affected	Hectare	966053	239511	674148	932111	3142685
Population affected	Number	38838	54263	755569	342567	634327
Human lives lost	Number	30	-	6	5	343
Cattles lost	Number	1998	142	429	431	6596
Villages affected	Number	509	127	680	756	1229
Area eroded	Hectare	-	748	429432	12539	7629
Value of crop loss	Rupees in lakh	17331	835	17432	14700	-
Value of houses damaged	Rupees in lakh	-	1648	259	4119	1869
Total value of damages	Rupees in lakh	_	19000	1976	18900	16590

Source : Water Resources Deparment, Government of Assam, 2006

However, in order to prevent the menace of recurring floods, different flood management schemes/projects were adopted in the state. They are generally the embankments, drainage channels, major sluices, anti-erosion work, etc. By adopting short term measures, the water resources department also protected land from flood and erosion in the flood prone areas. For effective management of flood and erosion in the state, long term measures in the form of storages reservoir in the upper catchments areas of the basin and watershed management were initiated. Artificial lakes, like Kurichu Hydel project at about 30 km upstream of Kurichu in Bhutan gave way to the retention of successive rain water. However, the artificial lake at Kurichu Hydel project sometimes created havoc downstream with inundation and evulsions. In Kopili Hydel project also excess water was released downstream. The embankment system overtopped resulting in heavy breaches and causing widespread devastation.

There are some environmental costs of irrigation developments, which should be avoided through proper planning and improvement of water management system all the way from the headwork to the tail end of the canal system to make irrigated agriculture sustainable. Since the main goal of irrigation development is to make agriculture more stable, productive and sustainable,[1] it is all the more important that the factors which reduce the efficiency of irrigated farming should be scientifically controlled. The impact of irrigation on productivity and sustainability of agriculture and environment can be realized only when water use is integrated with efficient soil, water, crop fertilizer and agronomic management system. Drainage is an essential component of irrigation system which has to be designed taking into consideration the soil and climate factors. More gains can be achieved from improving efficiency per unit of water. More gains are also expected from development of minor irrigation and in situ water harvesting and recycling than from building gigantic storage dams. Some of these are policy matters and others of technological and sociological nature.[2]

It may be feasible to harness the natural systems in bio-sphere through wet-land conservation and water-shed management as well as ground water recharge through the network of the water canals and rain water harvesting depending upon the geographical characteristics of specific sites, a combination of natural and artificial systems could be selected to meet the water needs of the local community. Scientific information on the assessment and monitoring of ground water resources in the north-eastern region, including Barak and Brahmaputra rivers, needs to be generated. Traditional way of water harvesting for the local community were important successful examples of water harnessing and to provide the aquatic ecosystem benefits to the communities. Water harvesting is suitable to the local climate and its storage is the key factor to adaptability to meet the water demand and the threat of climate change. To deal with the need of water storage arising out of climate change, it is necessary to prepare a working plan in a comprehensive manner[3] (Sharma : 2010).

The critical issue remains then how to store massive quantities of rain water falling in a very short period so that it can be used for the whole year. The issues of rain water management for strategic planning are (a) to retain as much water as possible in the soil profile, keeping in view the soil characteristics; (b) to recharge the ground water; (c) to manage the runoff to reduce erosion of soil and incidence of flash floods; (d) to store the inevitable runoff within the catchments; (e) to reduce stagnation of rain water in the cropped area by providing drainage; and (f) to increase productivity of cultivated lands, grasslands, bush and shrublands and forest and woodlands by extensive research for developing strategies for optimizing cropping pattern, water shed development and associated land treatment for diversification of cropping pattern.

Again, in a situation when there is water scarcity, efficiency of flood water management could be achieved in several ways, like (a) to retain precipitation in situ and minimize the run off; (b) to reduce evaporation in relation to transpiration; (c) to use drought resistant crops that fit in the rainfall patterns; (d) to recycle the run-off water after harvesting and drainage: and (e) to use water-shed concept for maximization of rainwater in situ. It is a combination of all these that ensure best results with minimum degradation of environment.

The Barak valley of Assam has a unique topography and it is endowed with abundant natural water resources. Apart from the river Barak and its tributaries, the ponds, oxbow lakes, temporary pools, *beels*, *haors* and other small acquatic systems can provide abundant scope for sustainable agriculture and fishery development. Oxbow lakes are remnants of the meandering flood plain rivers, which are physically isolated from the respective rivers and generally capture small relic drainages. These are present throughout the Barak valley. These are used as highly productive fisheries by fishermen. All these need care for sustainable management for the sustenance of human life locally and, in turn, globally.

Let us now make an attempt to reflect upon some of the plausible technical designs and management interventions to ensure an efficient and sustainable water management system in areas like the Barak valley. The irrigation system should be designed for protective, and not productive, purposes.[4] This calls for a balanced cropping pattern. Otherwise, being guided by the profit-maximization objectives, it may generate income inequality among the farmers through adoption of water intensive cropping pattern and commercialization of agriculture at the cost of the small and marginal farmers. Attention should be given to the achievement of inter-personal and inter-spatial equity. It is imperative to regulate the cropping pattern through control of water supply. As observed, in some areas in the Barak valley water becomes a common property during the flood situation and leads to a case for free-riding problem. Therefore, systematic management of common property necessitates interventions by an authority. This is because common property is largely being overused and hence rapidly depleted by the self-seekers or free-riders. The situation is known as 'The Tragedy of The Commons'.[55]

Quite often, farmers appear to be locked into their own variety of the Tragedy of the Commons. The abundant issue of water through gravity flow system and consequent flooding remove any option for a farmer to grow anything but paddy. Possibility of a change to a more productive, more profitable and more water sparing cropping system is practically difficult.

What is desirable is to try a potentially feasible cropping pattern on a limited area and then to promote collective decision to change.

Fishery development is undoubtedly very important for the economy of the region like the Barak valley. But the uncontrolled growth of capture fishery is likely to adversely affect other forms of aquatic bio-diversity, even though it may mitigate the livelihood requirements of the fisherfolk on a short term basis. It has been found that the sector specific management interventions are not able to take care of cross-cutting issues like maintenance of ecological integrity of aquatic eco-systems. Therefore, integrated water resource management attempts are recommended to address these problems by recognizing that different uses of water resources are interdependent (Gupta : 2010). It is particularly important in this region which faces problems of excess of water during some period and lack of water at some other timeframe. Due to poor irrigation facilities, agriculture suffers when rainfall is insufficient and irregular. In fact, this is the area where rain water management should be given priority for different reasons. It appears that the small and marginal farmers generally prefer field bunding. However, in the water-shed management programme, the only thing that has changed is the name. It is emphasized that the highest priority should be given to runoff storage to recharge the ground water, reduce the fury of flash floods and to provide for supplemental irrigation.

It appears that the Barak valley region at times gets flooded with water; while during other times, it suffers from water scarcity, both drinking and irrigation. This is further aggravated by the deforestation, unplanned urbanization, and the associated input of sewage and municipal wastes into the rivers, besides picnicking, quarrying activities, disposal of industrial effluents and other human activities. All these anthropogenic activities result in the impairment of water quality, habitat destruction and loss of bio-diversity in the riverine systems in the form of soil erosion, sand casting, sedimentation, etc. in the region which is one of the bio-diversity hotspots in the world. Only government action will not be sufficient to meet this challenge unless it is accompanied by proactive co-operation of the local people, because the predominant reasons for such state of affairs are the unplanned urbanization, human disturbances, and ignorance of the people regarding ecological services the rivers provide. For this, it is important to sensitize the people about the ecological services provided by the rivers and how the people themselves are responsible for environmental degradations. In the meantime, there should be a long term policy for the management of the river system by integrating ecological and social factors. Panchayats or local administrative bodies should have to be involved and given adequate resources to undertake grass root level pragmatic policies like the construction of series of reservoirs and other protective measures. It is also desirable to utilize the potential of indigenous knowledge of watershed management along with scientific knowledge to bridge the gaps concerning mechanisms of harnessing and storage of water successfully (Sharma: 2010).

Existing approach to irrigation is looked upon only as a technological input in production. But for all practical purposes, it is to be recognized that these technological considerations are required to be placed within the context of socio-economic relations and conditions of production and exchange. It is not the technological feasibility that alone matters. The creation, maintenance and utilization in totality are contingent upon these relations. Efforts should have to be made to evaluate the interaction between the irrigation technology and social relations. In effect, rainwater harvesting, proper management of existing resources, watershed development and community participation together can help the successful attainment of sustainable development of agriculture and upliftment of the socio-economic conditions of the people.

It has been observed that setting up of big dams at Tipaimukh and Subansiri as a solution to the problem may cause new problems, since these are located at the confluence of Indo-Burma, Indo-Malayan and Indo-Chinese Biodiversity hotspot zone which are characterized by the presence of a large number of plants and animal species. Precisely, it would create extensive detrimental downstream environmental impact. There is a maximum likelihood that once the dams are constructed, different species which are categorized as endangered and threatened may become extinct. Eventually, the impacts are likely to be serious on different areas, such as crop production, navigation, siltation, ecological imbalance, river pollution, extinction of river aquatic life, etc. In fact, social and environmental impact of the large dams on land submergence and people's resettlement remains an important issue in the Northeast, as it is anywhere else.

Water, being a finite resource, requires an appropriate management intervention. The water resource in Barak valley, in particular, is characterized by excess water during monsoons (and sometimes, in pre-monsoons) and scarcity during the lean months. This is a specific situation requiring reconciliation between seasonal variations and can be resolved only by appropriate management intervention. There is a need for comprehensive water resource management for its conservation and sustainable utilization, because sectoral priorities are often at conflicts with each other. In view of all these, how to utilize water in a sustainable manner by ensuring equitable and efficient allocation of the naturally available water in the valley at different timeframe has been a major concern. In the absence of well-articulated mechanism to ensure efficiency in water management, mal-distribution also tends to take place with all its attendant socio-economic and environmental consequences. The sustainable development is the only answer to the ills and apprehensions.

NOTES

1. A system can be considered sustainable if it ensures that today's economic development is not at the expense of tomorrow's development prospects (World Commission on Environment and Development 1987). There are certain aspects of sustainable agriculture. They are (a) meeting the changing needs of today and tomorrow; (b) economic viability of enhanced productivity level; (c) successful management of resources - internal, external, renewable or non-renewable; (d) maintenance, preferably enhancement, of quality of environment; and (e) conservation of natural capital.

2. Even though water is appropriated by the farmers as common property, the users themselves have to take care of different activities like construction of additional channels, maintenance of physical works within their jurisdiction, etc. It requires a good understanding amongst them systematically. This may be termed as 'co-operation for co-operative common property management'.

3. Irrigation water is an input used for agricultural production. It has certain characteristics. One, irrigation is transformed from a free good to an economic good and in the process as a public good, although there is no change in either its quality or morphological characteristics. Two, irrigation water is divisible for its users. Three, it is a leading input for technological progress in agriculture. Four, it helps to improve the permanent nature of land. Finally, it bears a strong relationship to social organization.

4. Many traditional societies have evolved over time codes of conduct which result in a sustainable use of these resources.

5. Quite often, farmers appear to be locked into their own variety of the Tragedy of the Commons. The abundant issue of water through gravity flow system and consequent flooding remove any option for a farmer to grow anything but paddy. Possibility of a change to a more productive, more profitable and more water sparing cropping system is practically difficult. What is desirable is to try a potentially feasible cropping pattern on a limited area and then to promote collective decision to change.

REFERENCES

Carruthers, Ian (1988), 'Irrigation under threat – a warning', *Asian Regional Symposium*, Vol. 11, Hydraulics Research, Walling Ford, England, p.39.

Gupta, Abhik (2010), 'Integrated water resource management: A Barak valley Perspective', paper presented in the symposium on 'Sustainable Development of Water Resources in Barak Valley' organized by IQAC, Assam University and INEIS (Unpublished).

Hardin, G. (1978), 'The Tragedy of the Commons', *Science*, Vol. 162, Washington D. C.

Misra, Sib Ranjan (2006), *Managing Canal Irrigation in India*, Concept Publishing House, New Delhi.

Sharma, G.D. (2010), 'Climate change and water security', the paper presented in the symposium on 'Sustainable Development of Water Resources in Barak Valley' organized by IQAC, Assam University, and INEIS (Unpublished).

Thavaraj, S.H.C. (1982), 'The Importance of integrating non-engineering assets in irrigation system design' in Donald, C., Taylor and Thomas H. Wickham (ed.), *Irrigation Policy and the Management of Irrigation Systems in South-east Asia*, The Agriculture Development Council Inc., Bangkok, p. 15.

World Commission on Environment and Development (1987), *Our Common Future*, Oxford University Press.

2

Integrated Water Resources Management: A Northeast India Perspective

Abhik Gupta

[**Abstract:** Water, being a finite resource, needs proper management interventions. The Water regime in Northeast India is characterized by excess water during monsoon (and sometimes in pre-monsoon as well) and scarcity during the lean months. The sectoral approach currently adopted for water resource management is not appropriate for conservation and sustainable utilization of water resource, as sectoral priorities are often at conflict with each other. Again, sector-specific management interventions are not able to take care of cross-cutting issues like maintenance of ecological integrity of aquatic ecosystems. Integrated water resources management (IWRM) attempts to address these problems by recognizing that the many different uses of water resource are interdependent. Besides briefly tracing the history of IWRM in a global as well as Indian context, the paper also discusses the present status of and possible approaches towards developing a sustainable IWRM model for Northeast India.]

Introduction

The earth appears blue when seen from space because of reflectance from the oceans that cover more than 70% of the planet's surface. Thus water gives planet earth its own unique 'signature', besides being a substance essential for life to sustain and propagate. It is also a versatile substance that has an extremely wide spectrum of use. It is needed for drinking, cooking and bathing by humans and domestic as well as wild animals; for other everyday uses by humans; and for irrigation, power generation, transport, navigation and manufacture of steel, paper, textiles, and almost all conceivable items. Biologically, it is the major constituent of all protoplasm; most plants and animals contain more than 60 % of water by volume. Physically, having a high specific heat, it can absorb large amount of heat energy before it gets hot itself; its property of remaining as a liquid in the temperature range of 0-100^0 C allows it to remain in the liquid state in most places on the earth; and its high surface tension makes it form droplets and waves, allows plants to absorb water and nutrients from their roots to the leaves, and blood to move through tiny vessels in animals. Chemically, it is the 'universal solvent' that is able to dissolve a large number of different chemical elements and compounds. Because of these properties water is used and abused for depositing sewage and industrial effluents and as a coolant in industrial and power generation processes including nuclear power. Even in the conceptual arena, it is perceived as a resource by water or power engineers; as a powerful weathering agent by geologists and geographers; as home and habitat of myriad biodiversity by ecologists; as an economic good by development economists; as a substance that has cradled and shaped the course of human civilization by historians; and as an inspiration for art, music and literature by all creative minds. These perceptions are not only different from one another; they often have a fair degree of 'conflicts of interests'. Again, while valuing water as a precious resource, one should also not overlook the numerous services that aquatic ecosystems provide 'free of cost'. To furnish a few examples, water detoxifies, purifies and cleanses plant and animal bodies and entire landscapes; the world ocean regulates climate and is the largest reservoir of carbon; unless polluted and contaminated by humans, freshwater ecosystems like ponds, lakes, springs, streams and rivers provide pure and safe drinking water (Pidwirny : 2006). However, only 3% of the total water is freshwater, the remaining 97% being saline water in the oceans. Of this 3%, around 99.7% is locked in ice caps and glaciers, in groundwater and in rocks; only 0.3% is there as surface water which in turn has the following distribution: 2% of all freshwater in rivers; 11% in swamps; and 87% in lakes (Gleick: 1996). Thus the freshwater reserve of the earth is finite and to add to this limitation it is unevenly distributed both in space and time – and shortage as well as excess - creating problems of management. The other major problem is the question of quality, which is being increasingly compromised by the entry of organic and inorganic contaminants. It is also important to remember the havoc-wreaking potential of water: water-borne diseases and water-related disasters are common sources of misery to a large number of people all over the world, but more so in the humid and underdeveloped tropics. A priority area of water management, therefore, comprises prevention and control of the frequency and magnitude of these episodes.

The conventional approach towards water resource management has been the supply-development approach, which in India is executed through the 'watertight' conduits of the 'Line Departments' of the state governments: flood control, irrigation, public health engineering and others. More recently, in recognition of the role played by aquatic ecosystems as repositories of biodiversity and in other ecological functions, the department of environment and forests is also involved in the identification, protection and conservation of biodiversity-rich areas of aquatic ecosystems. Furthermore, in keeping with the endorsement of the major principles of 'Integrated Water Resources Management (IWRM)' by the central government in several documents and in the National Water Policy, 2002 (Bhatt: 2004), many state governments like that of Assam have also adopted IWRM as a guiding principle in their state water policies. Accordingly, the erstwhile flood control department of the state has been renamed as the department of water resource in an attempt to inject holistic concepts into an essentially reductionist and 'embankment-bound' view of the riverine ecosystem. In spite of these changes, IWRM is still a nascent concept in India that faces numerous hurdles and challenges both in its adoption and implementation. It has often been suggested that the still largely informal water economy of India is not yet prepared for an all-pervasive incorporation of

IWRM (Shah and van Koppen : 2006), while others (Biswas : 2004; Saravanan : 2006) have held the inherent weaknesses of IWRM itself responsible for its inability to address the complex water issues of developing countries like India. This paper tries to assess the usefulness of IWRM in an Indian and more so a Northeast Indian context, at the same time providing a review of some of the major works that have analyzed and evaluated the theory and practice of IWRM.

A Brief History of IWRM

Integrated Water Resources Management (IWRM) has been defined by the Global Water Partnership (GWP) in 2000 as "*a process which promotes the co-ordinated development and management of water, land and related resources, in order to maximize the resultant economic and social welfare in an equitable manner without compromising the sustainability of vital ecosystems*". It is an avowed objective of the GWP to promote and apply the IWRM approach [http:// www. gwp.org/Global/About%20GWP/Strategic%20documents/ GWP_ Strategy_2009-2013_final.pdf]. The concept of IWRM in a comprehensive form is believed to have originated at the International Conference of Water and the Environment, at Dublin, Ireland, in January, 1992, a few months prior to the United Nations Conference on Environment and Development (UNCED) at Rio de Janeiro in June, 1992. The four guiding 'Dublin Principles' comprised a new approach towards the sustainable management of water resources:

1. "Freshwater is a finite and vulnerable resource, essential to sustain life, development and the environment"
2. "Water development and management should be based on a participatory approach, involving users, planners and policymakers at all levels"
3. "Women play a central part in the provision, management and safeguarding of water"
4. "Water has an economic value in all its competing uses and should be recognized as an economic good"

Comprising the 'heart' of IWRM, these principles had profound implications in policymaking in the water sector worldwide. For example, by admitting the first principle that water is essential for sustaining life, water ceased to remain an 'inert' resource that could be subjected to any form of engineering manipulation without taking into account the needs and requirements of the organisms living in or dependent on it. At the same time, the importance of water in bringing about development was recognized as well, thereby recommending a holistic approach that links development with conservation. The second principle brought down the decision making process to the lowest possible level, ensuring public consultations and involvement in water projects, while the third was perhaps among the first to acknowledge the "pivotal role of women as providers and users of water and guardians of the living environment". Finally, recognition of the economic value of water in the fourth principle was intended to achieve "efficient and equitable use", and put an end to "wasteful and environmentally damaging uses" of water (Snellen and Schrevel : 2004). Within this principle, it is important to first recognize the fundamental right of all human beings to have access to safe water and sanitation at affordable price [http://www.gwp.org /The-Challenge/ What-is-IWRM/Dublin-Rio-Principles/]. Biswas (2004) has, however, pointed out that the concept of IWRM was being promoted by agencies like UN since the 1950s, and integrated approaches were also recommended at the UN Water Conference at Mar del Plata, Argentina, in 1977. It is indeed true that although the nomenclature was not coined at that time, many elements of IWRM were present in the aims and objectives of the Tennessee Valley Authority (TVA) established in 1933. TVA integrated the demands of navigation, flood control and power generation, at the same time taking care of issues like erosion control, public health and others. Therefore, TVA can be said to have envisioned natural resource utilization from economic, social as well as environmental perspectives.

TVA had its impact on water management in India as well, when the Damodar Valley Corporation (DVC) was commissioned in 1948 as the first multipurpose river valley project of independent India. The objectives were flood control; power generation, transmission and distribution; irrigation; eco-conservation; afforestation and bringing about the social and economic well-being of the people living in the project catchments area. It is noteworthy that a multidisciplinary soil conservation department was established in the DVC as early as in 1949, thereby reflecting environmental concerns in the project objectives, an aspect hardly given any importance at that time. Nevertheless, the fact remains that though the concept of integration was emphasized in many UN documents and programmes in the 1950s, these essentially meant the development of support services within irrigated agriculture and not among irrigation and other water use sectors. The Mar del Plata Conference, however, pleaded for "real coordination among all bodies responsible for the investigation, development and management of water resources". However, it has also been criticized for strongly recommending expansion of irrigated agriculture for avoiding 'future famines', without expressing any concern about the environmental impacts of irrigated agriculture and its high water demand. The Conference, however, expressed extremely valid concerns regarding community access to safe water, pollution, and increased cooperation in sharing water resources (Snellen and Schrevel : 2004). In any case, IWRM was subsequently identified as one of the Programme Areas in Chapter 18 of Agenda 21 (Protection of the Quality and Supply of Freshwater Resources: Application of Integrated Approaches to the Development, Management and Use of Water Resources) at the United Nations Conference on Environment and Development (UNCED) at Rio de Janeiro, in June, 1992. This conference expressed grave concern that "The fragmentation of responsibilities for water resources development among sectoral agencies is proving, however, to be an even greater impediment to promoting integrated water management than had been anticipated. Effective implementation and coordination mechanisms are required" (Agenda 21 : 18.6). "Integrated water resources management is based on the perception of water as an integral part of the ecosystem, a natural resource and a social and economic good, whose quantity and quality determine the nature of its utilization. To this end, water resources have to be protected, taking into account the functioning of aquatic ecosystems and the perenniality of the resource, in order to satisfy and reconcile needs for water in human activities. In developing and using water resources, priority has to be given to the satisfaction of basic needs and the safeguarding of ecosystems. Beyond these requirements, however, water users should be charged appropriately" (Agenda 21 : 18.8). Further, IWRM, with the goal of integrating "land- and water-related aspects, should be carried out at the level of the catchments basin or sub-basin." IWRM advocated for a multisectoral approach to water resources management that incorporates technological, socio-economic, environmental and human health issues. Another important requirement was to take into account community needs and priorities while planning for a "sustainable and rational utilization, protection, conservation and management of water resources"; and to ensure public participation, including that of women, youth, indigenous people and local communities in "water management policy-making and decision-making". Finally, IWRM strived to "identify and strengthen or develop .... appropriate institutional, legal and financial mechanisms so that water policies could serve as a catalyst for sustainable social progress and economic growth" (Agenda 21 : 18.9) [http:// www.un.org/esa/ sustdev/ documents/ agenda21 /english/Agenda21.pdf].

In spite of this prominence given to IWRM principles in UNCED, it took another 10 years for IWRM to be formally adopted and ratified at the World Summit on Sustainable Development (WSSD) at Johannesburg, South Africa, in 2002. Article 26 of the Plan of Implementation of WSSD called upon all countries to formulate and adopt IWRM and water efficiency plans by 2005. Water efficiency plans are also an important part of IWRM, which in turn is a 'cyclic and long-term process'. Several countries have since taken up IWRM in different ways: through new water policies and planning; water and wastewater reforms, and so on, India being no exception. However, IWRM development pattern among different countries can vary because of their specific 'geographical, historical, cultural and economic' contexts. There is, therefore, neither a single model of IWRM development nor are there uniform priorities (Jønch-Clausen : 2004).

IWRM in India

Though many multipurpose river valley projects in India

might be said to have incorporated elements of IWRM, the first formal response came after the adoption of this approach in the WSSD at Johannesburg in 2002. Although India had already framed a National Water Policy (NWP) in 1987, 'a number of issues and challenges' had 'emerged in the development and management of the water resources'. This necessitated a review and updating of the policy (NWP : 1.10), which was finally adopted by the National Water Resources Council in its 5th meeting held on April 1, 2002. The policy document does not mention IWRM *per se*, but states that "Water is a scarce and precious national resource to be planned, developed, conserved and managed as such, and on an integrated and environmentally sound basis, keeping in view the socio-economic aspects and needs of the States. It is one of the most crucial elements in developmental planning" (NWP : 1.4) ... and ... "there should be an integrated and multi-disciplinary approach to the planning, formulation, clearance and implementation of projects" (NWP : 6.4). The Policy recommends adherence to an important IWRM principle by stating "Water resources development and management will have to be planned for a hydrological unit such as drainage basin as a whole or for a sub-basin, multi-sectorally, taking into account surface and ground water for sustainable use incorporating quantity and quality aspects as well as environmental considerations .... Watershed management through extensive soil conservation, catchments-area treatment, preservation of forests and increasing the forest cover and the construction of check-dams should be promoted. Efforts shall be to conserve the water in the catchments" (NWP : 3.3, 3.4). The Policy also prioritizes water allocation as: (i) drinking water; (ii) irrigation; (iii) hydro-power; (iv) ecology (v) agro-industries and non-agricultural industries; and (vi) navigation and other uses (NWP : 5) (National Water Policy : 2002).

While IWRM principles have been recognized in the water policy, to what extent has it been possible to operationalize these principles? In a series of studies (Grizzetti *et al.* : 2008; Nagothu and Manasi : 2008; Manasi *et al.* : 2009) conducted under the STRIVER (Strategy and Methodology for Improved IWRM) programme in the Tungabhadra Sub Basin (TBSB), which lies in the two southern states of Karnataka and Andhra Pradesh, it was observed that IWRM principles were found to be somewhat incorporated in a few instances which were only on a local scale, while water management at large was still based on administrative and not hydrological boundaries. Irrigation accounted for about 94% of the water use, and this along with effluents from mining, industrial and domestic sources comprised the main source of anthropogenic disturbances leading to ecological disruption in the river. Further complications were caused by the trans-boundary nature of the river and the conflict between the two neighbouring states over the sharing of the river water. However, there had been several attempts in the recent years to improve stakeholder participation which was one of the cornerstones of IWRM. The policy support for this was provided by the National Water Policy, 2002 and Karnataka State Water Policy, 2002, with a participatory approach also having been espoused by the Andhra Pradesh Farmers' Management of Irrigation Systems Act, 1997. Similar participation was also taking place in the drinking water and watershed management sectors. Nevertheless, the fact remains that these initiatives in integration were still at a local level, and up-scaling strategies were yet to be fully operationalized. Considerable progress was made in the watershed protection sector by improving forest cover as well as by declaring large areas in TBSB as National Parks and other Protected Areas. With regard to maintenance of water quality, besides regular monitoring by the Karnataka Pollution Control Board, the civil society was playing an increasingly pro-active role in influencing the industrial sector to treat effluents and reduce water pollution by forming the Tungabhadra Watchdog Committee.

The Northeast Indian Scenario vis-a-vis IWRM

The eight Northeast Indian states of Arunachal Pradesh, Assam, Manipur, Meghalaya, Mizoram, Nagaland, Sikkim and Tripura largely belong to the Meghna (Barak) and Brahmaputra Sub Basins of the Ganga-Meghna-Brahmaputra Basin, barring a small area under the Chindwin-Irrawaddy Basin. The main river of Sikkim, the Teesta, also drains into the Brahmaputra in Bangladesh. The combined average annual flow of the Brahmaputra and the Meghna (Barak) Sub Basins is 677.41

cubic kilometre (km^3=billion cubic metre or BCM), while another 8.57 km^3 comes from rivers flowing directly into Bangladesh and 22.43 km^3 from rivers flowing into Myanmar. Of this enormous flow, however, the utilizable flow is only about 24 km^3 (Kumar *et al.* : 2005). The total replenishable groundwater resources in the region is estimated to be between 31.24 and 34.24 km^3 (Kumar *et al.* : 2005; Garg and Hassan : 2007). Taking into account a live storage capacity of 67.37 km^3 for surface water, the total utilizable surface water amounts to 59.014 km^3 and groundwater 35.165 km^3, making the total water resources to be 94.179 km^3. With a projected requirement of 40.95 km^3 in 2050, the Brahmaputra-Barak Basin remains a water surplus basin as compared to several deficit basins such as the Indus, the Ganga, the Mahanadi, the Godavari, the Cauvery and others (Garg and Hassan : 2007). The per capita water availability in this region is 18,400 cu m against the all India average of 2208 cu m (Mahanta : 2006). The region has a hydropower potential of 35000 MW at 60% load factor, thereby accounting for 41.6 % of India's hydropower potential.

In spite of these positive attributes, Northeast India faces a plethora of water-related problems. Among the most pressing of these are those of floods in the valleys and landslides generated by heavy rainfall in the hills. The region is also riddled by the paradox of excess water in the monsoon months of June-September and scarcity in the dry period, especially during December-March. While flooding by excess water is common, there are drought years as well. The dry season and/or drought year scarcity affects both quantity and quality of drinking water supply in many areas, as in the absence of flushing and replenishment, pollutant build-up affects surface water sources like rivers, streams, ponds and tanks. Poor drainage in many urban centres such as Guwahati, Silchar and others lead to prolonged waterlogging and the resultant inundation and contamination of groundwater sources. Agriculture is also affected, as the *boro* paddy and other *ravi* crops that are cultivated in extensive floodplain areas suffer either from water scarcity or from flooding, often leading to distress sale/lease of prime agricultural land to brick kiln owners, resulting in irreversible destruction of top soil and defacement of landscape. The *kharif* crop is similarly affected, more commonly by flood, but less frequently by water scarcity in drought years (Gupta, 2006). Although only 3.37 and 3.97 per cent of the groundwater potential of the Brahmaputra and Barak valleys, respectively, have been utilized so far (Verghese : 2006), the emerging threat of arsenic and fluoride contamination of groundwater is a major constraint on the use of groundwater for drinking and agricultural water supply in Brahmaputra, Barak and Manipur valleys (Singh, 2004; Chakraborti *et al.*, 2004).

It may be worthwhile to examine the most contentious and hotly debated issue of hydropower generation in the region from an IWRM perspective. Both electricity generation and consumption are low in the north-eastern states, which have a total hydropower potential of about 58,971 MW (about 35000 MW at 60% load factor), although only around 2 per cent of this potential has been utilized till today [http://www.projectsmonitor.com/detailnews.asp?newsid= 17734]. The Government of India had taken up the "50000 MW Hydroelectric Initiative" in 2003 for augmenting hydropower generation in the country. Under this initiative, preliminary feasibility reports (PFRs) of 162 hydro-electric projects were to be prepared to achieve a total power generation of 47930 MW installed capacity, of which the share of the north-eastern states is 31885 MW or about 66.5 per cent of the total capacity building. An installed capacity of 27293 MW (85.6 %) of the latter is proposed to be developed in Arunachal Pradesh alone. This staggering development includes mega projects like Dibang Multipurpose Project with an installed capacity of 3000 MW, Lower Siang (2700 MW), Upper Siang-Siang Intermediate (11000 MW), Demwe Lower (1750 MW), Demwe Upper (1800 MW), Subansiri Middle (1600 MW), and others. Work on Lower Subansiri (2000 MW) has already started [http://www.cea.nic.in/]. Meanwhile, there have been strong protests and agitations in Assam, which is the lower riparian state, against the setting up of so many large dams in the upstream state of Arunachal Pradesh. The 1500 MW Tipaimukih project on River Barak had to be shelved in the backdrop of protests in Manipur as well as in the lower riparian country of Bangladesh. There are three major concerns about the creation of dams in Northeast India, the first being the risk of dam failure due to earthquakes or heavy rainfall events, as the region is located in the highly seismic Zone V and also receives heavy rainfall during the monsoon. The other concerns include the destruction of forests, loss of biodiversity and indigenous religious-cultural relics and heritage in this global biodiversity hotspot, as well as the vexed issue of relief and rehabilitation of displaced persons. The power ministry, while acknowledging the inadequacy of mitigation measures adopted, has made it mandatory for project developers, especially public sector companies, to undertake afforestation as a part of project activities [http://chimalaya.org/category/hydropower-projects-and-climate-change/].

Nevertheless, it appears from the outcome of different public meetings, seminars and conferences that public opinion in Assam at large is against the construction of large dams in this region, and imposing such decisions go against the principles of IWRM, which advocates for decisions to be "taken at the lowest appropriate level, with full public consultation and involvement of users in the planning and implementation of water projects" (Snellen and Schrevel : 2004). This calls for extensive public consultations to be initiated in all the project-affected areas of Arunachal Pradesh and Assam, before taking any decisions on these projects. However, of the 160-odd hydroelectric power projects identified by the Central Electricity Authority under the 50,000 MW Hydroelectric initiative, about 42 projects between the installed capacity of 30-100 MW can yield a total of 2747 MW, of which some 32 projects with a total installed capacity of 2198 MW appear to be largely free from controversies and major adverse impacts. Additionally, at least about 612 MW can be generated from small hydropower projects. Thus a total installed capacity of 2810 MW can come from such small and medium-scale projects that are likely to have less environmental and social impacts. These additions themselves will more than treble the present generation of 1116 MW in the region. Therefore, these projects could be examined with priority and executed if found feasible, while more extensive studies should continue to be conducted on the controversial mega projects before arriving at any decision with informed consent of all stakeholders accorded uppermost priority. An atmosphere of trust and transparency has to be created among project developers and project affected population. Furthermore, smaller projects are likely to have shorter gestation periods and begin to address the existing power shortage in the region within a relatively short time. In many cases, projects could be downsized to arrive at an acceptable solution. For instance, the 600 MW Kameng Hydroelectric Project has been reviewed to an installed capacity of 480 MW to avoid submergence of Seppa town in Arunachal Pradesh. Similar reviews could be conducted on large projects with major environmental and socio-cultural impacts as well as high seismicity-related risks. The aim should be to arrive at not only informed consent of stakeholders at all possible levels, but also informed choice of the different technological options available, as emphasized in IWRM, as well as in various UN directives. Institutional reforms such as replacement of the Brahmaputra Board by a North- Eastern Water Resources Authority (NEWRA) attached to the North-Eastern Council (NEC), and with powers to issue environmental and other clearances to all regional projects (Verghese : 2006) have also been proposed. This would also enable addressing issues in a more holistic manner in view of the highly integrated and interactive nature of the highland-lowland ecological systems existing in Northeast India where any isolated and unilateral environmental alteration in the highlands could lead to hazards to the detriment of both the hill and the valley people and ecosystems. Viewing the options in the larger context of the Brahmaputra-Barak-Meghna Sub-basin of the Ganga-Brahmaputra-Meghna Basin and including interstate and intercountry dialogues among the hill states and Assam as well as among Bhutan, India and Bangladesh could lead to a much better and environmentally, socially and culturally acceptable and sustainable utilization of the huge water resources of this region of our country. Application of IWRM principles could go a long way in successfully addressing these so far contentious issues.

However, a number of grey areas still remain even in the institutional and policy mechanisms. For example, Assam is the only state to formulate a State Water Policy largely along the lines of the National Water Policy, 2002. The Assam policy unequivocally declares IWRM as the core strategy and aims for developing a "systematic knowledge base to examine the current understanding of integrated freshwater management quantitatively and qualitatively." The Policy has prioritized drinking water and water for domestic use as the most important water use followed by irrigation water; hydropower with a caveat of "within the constraints imposed by others"; industrial use including agro-based industries; and finally "for recreation, health, navigation and other uses" [http://www.old.indiawaterportal.org/data/policies/ StateWaterPolicy Assam_Draft.pdf]. One significant lacuna appears to be the low priority given to the value of water as habitat for wildlife and its other ecological values, especially in a state like Assam which has R. Brahmaputra and the associated wetlands with their rich, unique and yet not fully explored biodiversity. In this regard, the State Policy also deviates from the national policy which accords relatively high priority to ecology, albeit at the fourth place after drinking water, irrigation and hydropower. Paradoxically Assam is one of the few Indian states that has shown due regard for its aquatic biodiversity by selecting the Gangetic Dolphin as the State Aquatic Animal, thereby giving it a flagship status. This incongruity probably points out the lack of coordination and information sharing among the departments, which remains one of the major hurdles towards adopting the principles of IWRM at the departmental level. Another important strategy in Assam State Water Policy is including user participation and decentralization of authority. Several NGOs and citizens' groups have apprehended that decentralization might become a synonym for rampant privatization of water services. They also called for proper regulation of the 'commoditization' of water that is likely to follow the decentralization process, thereby robbing the communities of their traditional rights and privileges and instead becoming a burden on their frugal economy [http://www.sandrp.in/wtrsect/Assam_Water_Policy_ Concerns_of_RVC_Assam_Aug2007.pdf]. Therefore, these concerns as well as the growing public opinion against indiscriminate construction of large dams without conducting extensive public hearings in downstream areas reflect the increasing stakeholder awareness and participation in water-related decision making in Assam.

Among the other Northeast Indian states, Nagaland also proposes to draft a state water policy, as despite having abundant rainfall in the monsoon, the state suffers from water scarcity in the lean months [http://www. zeenews.com/news636222.html]. Meghalaya has created a separate water resources department in 2009, thereby taking a step towards recognition of IWRM principles [http://www.livemint.com/2009/07/30164817/Meghalaya-creates-water-resour.html], while Arunachal Pradesh has enacted "The Arunachal Pradesh Water Resources Regulatory Authority Act" in 2006 to establish an Arunachal Pradesh Water Resources Regulatory Authority. A major task of this body is to determine the distribution of water for different categories of uses. It is also to ensure equitable distribution of water during periods of scarcity and to establish a water tariff system, working within the framework of the state water policy (Anonymous, 2006). Thus there is lack of a coordinated approach among the different states in their water resources management strategies, and although some elements of IWRM can be said to have been included or at least recognized, incorporation of IWRM in the actual practices is still a far cry. Often the establishment of an independent water resources department or authority could simply be to better avail the benefits from centre-sponsored schemes. There is clearly scope for achieving policy uniformity among states, consensus on conflict issues and improvement of governance. Nevertheless, certain crucial tasks such as the cleaning of rivers in the coal belt of Meghalaya could now be initiated with central assistance.

Hence, notwithstanding the limitations, it might be admitted that acceptance of IWRM at the conceptual level has led to some positive and holistic steps in water resources planning and management, in spite of the criticisms of IWRM levelled by scholars like Saravanan (2006), among others. Biswas (2004) has also opined that it would be extremely difficult to operationalize the principles of IWRM, in spite of its popular acceptance in many countries across the globe. However, it must be remembered that as the water problems and issues are becoming increasingly complex with multi-sectoral ramifications, more so with climate change and the global food crisis looming large as new challenges, no single reductionist approach could possibly address them with any amount of success. This further underscores the need for a holistic and integrative approach like the IWRM that aims to strike a balance among myriad competing stakeholder requirements.

IWRM is a philosophy and not a straight-jacketed action manual and converting philosophy into action always remains a challenge, especially in largely informal water economies like that of India, a constraint that has also been pointed out by the International Water Management Institute (IWMI : 2007). In view of the fragile nature of freshwater ecosystems in Northeast India that harbour a rich instream as well as riparian biodiversity, the informal and traditional nature of water use in most indigenous societies, and the conflicting demands on water allocation in different sectors that is likely to intensify in the coming years because of the water surplus nature of the Brahmaputra-Barak Sub-basin, the integrating and moderating influence of IWRM could be a powerful tool in evolving a sustainable and socially acceptable water use model in this region of our country. Snellen and Schrevel (2004) have rightly pronounced "It has also become clear, from the very start, that progress is difficult and slow. Indeed the operational questions are complex. However, it is generally accepted that to manage water resources there is no alternative to IWRM."

REFERENCES

Anonymous. 2006, 'The Arunachal Pradesh Water Resources Regulatory Authority Act, 2006 (Act No. 15 of 2006)', *The Arunachal Pradesh Gazette*, No. 106, Vol. XIV, December 15, 2006, Naharlagun. Government of Arunachal Pradesh, Law and Judicial Department, Itanagar.

Biswas, A.K. (2004), 'Integrated Water Resources Management (IWRM): A Reassessment', *Water International* 29 (2): 248-256.

Bhatt, S. (2004), 'Scoping exercise for mainstreaming the ecosystem approach into integrated water resource management, India', IUCN Water and Nature Initiative: Final Report. [http://app.iucn.org/dbtw-wpd/edocs/Rep-2004-018.pdf.].

Chakraborti, D., Singh, E.J.K., Das, B., Shah, B.A., Hossain, M.A., Nayak, B., Ahmed, S. and Singh, N.R. (2009), 'Groundwater arsenic contamination in Manipur, one of the seven North-Eastern Hill states of India: a future danger', *Environ Geol.* 56: 381-390.

Garg, N.K. and Hassan, Q. 2007, 'Alarming scarcity of water in India'. *Current Science* 93 (7): 932-941.

Gleick, P. H. (1996), 'Water resources', In: *Encyclopedia of Climate and Weather*, ed. by S. H. Schneider, Oxford University Press, New York, vol. 2, pp. 817-823.

Grizzetti, B., Lo Porto, A., Barkved, L.J., Bouraoui, F., Deelstra, J. and Joy, K.J. (2008), 'Modelling water pollution with stakeholders involvement – The twinned experience of Glomma (Norway) and Tungabhadra (India) river basins' Striver Policy Brief No. 10. [www.striver.no.].

Gupta, A. (2006), 'Water Availability, Poverty and Socio-Economic Crisis in the Floodplains of Barak Valley, Assam, North East India' [https://www.mpl.ird.fr/ UR044/projets/ Textes/ Gupta. doc.].

IWMI, 2007, 'IWRM challenges in developing countries : lessons from India and elsewhere', *Water Policy Briefing – Issue 24*, International Water Management Institute [http://www. iwmi.cgiar.org/ Publications/ Water_Policy_Briefs/ PDF/ WPB24.pdf].

Jønch-Clausen, T. (2004), *Integrated Water Resources Management (IWRM) and Water Efficiency by 2005: Why, What and How?*, Global Water Partnership, Sweden.

Kumar, R., Singh, R.D. and Sharma, K.D. (2005), 'Water resources of India', *Current Science* 89 (5): 794-811.

Manasi, S., Latha, N., Paranjape, S., Joy, K.J., Nagothu, U.S., Raju, K.V. and Mollinga, P.P. (2009), 'Strategies and recommendations towards an IWRM approach in Tungabhadra sub-basin, India', Striver Policy Brief No. 1. [www.striver.no.].

Mahanta, C. (2006), 'Water Resources of the North East : State of the knowledge base', Background Paper No. 2, Draft for Discussion – March 2006. [http://mdoner.gov.in/ writereaddata/ sublink3images/ KnowledgeBase 3697746260.pdf].

Nagothu, U.S. and Manasi, S. (2008), 'IWRM and livelihoods: fisheries in Tungabhadra sub-basin, India, Striver Policy Brief No. 1.[www.striver.no.].

National Water Policy (2002), Ministry of Water Resources, Government of India, New Delhi.

Pidwirny, M. (2006), *Fundamentals of Physical Geography, 2nd Edition*. [http://www. physicalgeography.net/ fundamentals/ contents.html.].

Saravanan, V.S. (2006), 'Integrated Water Resource Management: A Response', *Economic and Political Weekly*, September 23: 4086-4087.

Shah, T. and van Koppen, B. (2006), 'Is India ripe for integrated water resources management: fitting water policy to national development context', *Economic and Political Weekly* 41 (31): 3413-3421.

Singh, A.K. (2004), 'Arsenic contamination in the groundwater of North Eastern India', In: *Proceedings of National Seminar on Hydrology*, Roorkee.

Snellen, W.B. and Schrevel, A. (2004), 'IWRM: for sustainable use of water – 50 years of international experience with the concept of integrated water management', *Background document to the FAO/Netherlands Conference on Water for Food and Ecosystems*, Ministry of Agriculture, Nature and Food Quality, The Netherlands. [http://www.fao.org/ ag/wfe2005/ docs/ IWRM _ Background.pdf]

Verghese, B.G. (2006), 'Water resources in the Northeast: Development options in a cooperative framework', Background Paper No. 1, Draft for Discussion – March 2006. [http://mdoner.gov.in/ writereaddata/ sublink3 images/cooperation 8437515003.pdf].

3

Climate Change and Water Resource Management in Northeast India

G.D. Sharma

[Abstract: In the month of December, 2009, the heads of various countries met in Copenhagen (Sweden) to discuss the problems associated with global warming. The meeting was necessitated by the atmospheric changes causing threats to life in the planet. The anthropogenic activities clubbed with the use of fuel energy for various developmental activities are adding carbon-di-oxide and nitrous oxide gases to the atmosphere in huge amount. The whole world is experiencing the general warming trend of the earth. The global average temperature has risen to 0.74^0 C over the last 100 years. Based on the various climatic models, it can be predicted that the temperature on the Indian subcontinent will rise between 3.5 and 5.5^0 C by 2100. This change in atmospheric temperature is directly related to the changes in the dates of the onset and retreat of the monsoon as well as the number of rainfall days and maximum changes in its frequency. The effect of global warming on Northeast India was visible by unprecedented draught in some states of the region and floods in other states during the years 2008-2010.]

Introduction

Water is not only an important resource; it is the lifeline of human existence. All living beings need water for their cellular activities. The life forms and diversity on the earth planet are determined by the quality and quantity of water. The water cycle determines the pattern of water systems which constitute 4% of world's water resources. Water resource has to support the world's 15% human population and 20% livestock and agriculture practices. The rapid growing population, industrialization, deforestation and improvement of quality of life are causing pressures on availability of potable water. Rainfall constitutes the vital asset for agriculture in India. The spatial and temporal distribution of water determines availability of water. The rainfall in the Northeast shows spatial and temporal variability from 11000 mm to 13000 mm per annum in the Cherrapunji area of Khasi Hills (Meghalaya) and 800-1200 mm in south-central Assam (Nagaon). There has been a significant decrease in rains in the region (Das, 2004). Due to variation in precipitation and population pressure, there has been over exploitation of ground water on one hand and heavy monsoon downpour which causes floods, landslides and erosions on the other hand. This has created a feeling in many that the North-Eastern Region (NER) is surplus in water and rainfall and therefore, conservation of natural water bodies is not necessary. This is not true, because the variability of rainfall in the region is very high in space and time.

Water Resources in Northeast India

In Northeast India, the rivers originate in the hills of trans- and sub-Himalayas. The river Brahmaputra in the northern and Barak (Meghna) in the southern parts of Assam are the major rivers. These two river systems are fed by numerous tributaries. Both the rivers pass through the Northeast and merge in Bangladesh to confluence into the Bay of Bengal. The Barak originates at the Manipur-Myanmar border, passes through South Assam and merges with Brahmaputra in Bangladesh, and thus the two river systems of the region obtain an international character. The surface water resources from these rivers amount to 652.8 billion cubic meters, i.e., 34% share of total Indian water resources. Mayhem is caused almost every year from annual floods from these rivers in Assam. Nearly 40% of land in the state is prone to floods which make a fragile agro-economic environment in the region. Floods are caused by combination of natural and anthropogenic activities. The landslides and heavy erosions during monsoon in the region is due to the fragile geo-environmental setting of the Eastern Himalayas, active seismicity, massive deforestation, intense land use pressure, high population growth and adhoc flood control measures. The floods of 1988, 1998, 2004, 2007 were worst exacerbated by the socio-economic condition of the people (Goswami & Das, 2002). To overcome such natural disasters, integrated water resources management involving traditional and modern technology, such as, embankments, spurs, dykes, bamboo crib, along with reviving of watershed and wetlands for multipurpose uses, is urgently called for. Massive afforestation, soil conservation, flood forecasting, awareness programmes to enhance capacity building, etc, are also necessary. This will help in harnessing the hydro-power potential of the region, nearly 40% of the country's (32000 mw) power requirement for development.

NER is rich in water resources, whereas flood is a perennial problem in the region. It has been observed over the years that the flood toll is rising in the region in every subsequent flood. More than 250 lives were lost in the last decade and the damage to crops, houses and public utilities was estimated at Rs. 19150 million. Three master plans for Brahmaputra, Barak and major tributaries were prepared, but these are found to be unacceptable in terms of technology and fund requirement. Mega projects are suggested to generate electricity and to reduce peak flood, but such projects are likely to cause large scale submergence of villages and displacement of people. The ecological consequences of such mega-electricity projects should also be properly studied before launching the projects. The river linking, proposed in some quarters, will have disastrous effect on the ecology of the region. The dam project on the Barak river at Tipaimukh in the tri-junction of Assam, Manipur and Mizoram will also cause submergence and displacement of a large number of tribal families (Hmar and Zeliangrong) and loss of bio-diversity and cultural wealth. On the other hand, the flood forecast can be improved by using advanced technology based on satellite data (by ISRO). The regional cooperation between India, Bangladesh, Myanmar, China, Nepal and Bhutan on sharing and management of water resources of NER can be more gainful to all the stakeholders. The North-Eastern Council (NEC) and ministry for the Development of North-Eastern Region (DONER) need to formulate a perspective plan for water resource management keeping in view the threat of climate change. NER is already challenged for a long time with floods and erosions and the delay in tackling the problems will result in the demand for additional funds which could be utilized for infrastructure development, employment generation and improvement of quality of life. The shortage of drinking water facilities in the cities is related to inadequate fund allocations and inappropriate distribution system. The scarcity is man-made and largely due to lack of managerial foresight.

Access to Water

Water pollution affects the poor people who do not get access to drinking water. They suffer with a number of water borne diseases like cholera, diarrhoea, hepatitis etc. The quality of water determines not only the survival, it is also essential for many livelihood related activities. The cost of access to quality water deters the people's capability to work. This also determines the overall economic development of the region. In spite of plenty of water during rainy season in the Northeast, the sustainable access in dry period still remains a dream. One of the major factors for low agricultural productivity and aquaculture in the Northeast is general, and Barak Valley in particular, is access to water during winter and spring seasons. The government has many strategies to reduce poverty. However, appropriate budget allocation is needed to address the multiple causes of poverty. Integrated strategies like local infrastructure development for harnessing rainwater, water supply and sanitation, participation of people and good governance are important for improvement of the quality of life of the poor. Public-private-partnership can help the poor better access to water and sanitation.

Traditional Institutions

The tribal people of hill areas in the Northeast have some traditional laws and customs on the use of natural resources, including water. The purity of natural streams and conservation of water resources are done effectively by those traditional institutions. In the plain areas, however, such institutions are not effective. On the other hand, institutional arrangements for irrigation and water management is becoming more and more necessary for water related infrastructure in the region. The landscape of the Northeast is dominated by natural wetlands that include natural lakes as well as natural and man-made tanks and ponds. Mythological beliefs in many cases booast the supernatural origins of these water bodies and their conservation by the deities. These water bodies integrate culture and traditional wisdom to harness rain water and replenish ground water table. Ministry of Water Resources, Government of India, has initiated several schemes for water conservation by involving the local people who are the users and can be the best conservators of the natural water bodies in their respective localities. Involving the traditional institutions in those schemes may also be highly rewarding.

The increasing deforestation and pressure of population coupled with expanding urbanization have reduced the number of natural water bodies and slackened the control of traditional laws and customs over them. A number of aquatic species have been lost in the process. Pattern of migratory birds has also changed significantly. In last 50 years, many small water bodies have disappeared, thereby affecting many socio-economic activities in the region, specially the aquaculture and the prospects of eco-tourism. Revival of water bodies is, therefore, necessary for the conservation of eco-systems as well as the sustainable economic growth of the region. The water bodies which help to sustain native flora and fauna should be maintained by involving the local communities as economic and ecological resources. This can be achieved by reviving the indigenous knowledge and traditional methods of water management.

Traditional Management of Water Resources

Northeast India is rich in cultural diversity and is a hotspot for bio-diversity. The people in the region have preserved their natural resources for ages as part of their cultural heritage. Collective action of the people is a powerful instrument for the management of all natural resources. The village communities used to make embankments for rain water harvesting and the purity of the harvested water was ensured by customary rites and practices against pollution and contamination. The harvested water could thus be used for drinking as well as all other human activities. The tribal communities of the region had successfully developed indigenous technology for harvesting, preserving and conserving rain water. The traditional customs and rules are the local regulatory systems to maintain and manage the commonly used water bodies. (Ao, 2010) The Khasi and Jaintia tribes of Meghalaya, Apatani and Nishi tribes of Arunachal Pradesh, Chakesang tribe of Nagaland had developed water harvesting both for irrigation and drinking on steep hills of their respective areas. The Bodo tribe of Assam developed indigenous canal system to cultivate wet rice in very early period of the history of their settlements. These man-made streams, called *dang*, were canalized from distant hill streams or lakes to the rice fields. Well developed canal (*Dong*) system could irrigate from 1000 to 5000 hectares of land in the dry season. Such collective enterprise could meet the need of villages and help in the conservation of water resources. Another age old practice of water harvesting in the plains of Assam was the construction of water tank or pond (*pukhari*). Such ponds were meant for common public use. Besides the religious beliefs and cultural values, these ponds also served the economic and agricultural needs of people in the localities around them (Basu, 1970). However, the modernity has not continued the required support to these traditions. On the other hand, the anthropogenic factors like population pressure and urbanization have destroyed many of those traditional wetlands.

Climate Change and Water Resources

Climate change is generally referred to the rising of temperature, increased draught, uneven rainfall, unpredictable rainfall pattern, etc. These sets of weather conditions of a region maintain stable climate for longer durations without much variations, while the climate change causes significant variation or shift in the environment. However, in the context of water resources the climate change is mainly referred to the induced changes due to human activities. Due to human induced climate change certain visible parameters of weather, especially the fluctuations in intensity of rainfall and temperature, are becoming conspicuous. The exceptional rise of temperature up to 34°c in Shillong in the month of July 2010 and up to 39°c in the Cachar district in the month of August 2006 are examples of the impact of global warming in the Northeast.

One of the most serious concerns of climate change is about the availability of water. This problem, together with increased demand from a growing population of water for agricultural, industrial and domestic purposes, is likely to turn water into a major issue in the countries of Asia in the coming years. The problem of growing gap between demand and availability of water was discussed during 5th World Water Forum in Istanbul in March 2009. To deal with the problem, three major recommendations were made,

1) The need to increase the awareness and scientific information.

2) The need to develop natural and artificial systems to increase the water storage capacity.

3) The need to manage the water and rain water harvesting as the key issues for improving water security and adaptation strategy to the climate change due to global warming.

The problem has been aggravated by both natural and human factors. The glaciers are melting very fast due to rising temperature, where as the wetlands are facing threats by the increased exploitations towards land filling, either for human habitation or for agriculture. In fact, the Northeast represents a scenario of contrasts. The continuation of the traditional ways of water harvesting by Nishi tribe of Arunachal Pradesh and 'Zabo' system of cultivation by Naga tribes are good examples of watershed management. On the other hand, wetlands like Chatla hoar and Sonbeel in the Barak Valley of South Assam are facing threats of decay and extinction in parts from the over-exploitation by the growing population.

In the Asian scenario the increasing water scarcity is a serious threat arising due to climate change. It can be anticipated that the water availability will shortly approach the critical stress level in countries like Pakistan, Afghanistan, China and India. The global warming has also effected visible change on the pattern of rainfall; in some areas it is more than usual while in other areas the scarcity is leading to draught like situations. Rain, instead of snow, on hills is reducing both long and short term water storage in ice forms. Rains have also increased in intensity for shorter periods resulting in frequency of floods in higher proportions. There is also reduction in ground water recharge. This is leading to less storage of water for drinking and irrigation of farmland. The best example of skewed rainfall is Cherapunji in Meghalaya, which receives highest rainfall in India (about 12,000 mm annually). Of the total, nearly 50% rainfall is received in just about 15 days, whereas the same Cherajunji area faces shortage of water during winter months due to storage problems. The critical issue, therefore, is how to store massive quantities of rain water falling in a very short period so that it can be used for the whole year. Deforestation on hills and sedimentation of rivers and other natural water reservoirs is another great problem. Though large dams have been built for irrigation and producing hydro-electricity in many places, the problem of sedimentation and the social and environmental impact of the dams like submergence of land and displacement of people are important issues to be dwelt upon.

It is possible to utilise the potential of indigenous knowledge of watershed management along with scientific knowledge to bridge the gaps concerning mechanism of harnessing and storage of water successfully. It may also be feasible to harness the natural systems in biosphere through wetlands conservation and watershed management as well as ground water recharge through a network of water canals and rain water harvesting devises depending upon the geographical characteristics of the specific sites. A combination of natural and artificial systems can be selected to meet the water needs of the local communities. It is necessary to have the options of including the ponds, lakes and aquifers in a planned scheme of active water storage, rather than considering those as passive sources. Scientific information on the assessment and monitoring of ground water resources in the region, including Barak and Brahmaputra rivers, need to be generated. Traditional methods of water harvesting of the local communities are important examples of water harnessing. These methods may be scientifically improved upon to provide the aquatic ecosystem benefits to the local communities. Water harvesting is suitable to the local climate and its storage is the key factor to adaptability to meet the water demand and the challenges of climate change. To deal with the problem of water storage arising due to climate change, it is necessary to prepare a workable master plan for the entire Northeast in a comprehensive manner. It will require a strong political will and involvement of government, non-government agencies and the local communities to conserve the water resources and to ensure the availability of water to the future generations. Water being a precious liquid for all life related activities and health of the living beings, its availability and sustainability has to be guaranteed by the involvement and active participation of each one in the society. It has to be remembered that the most crucial issue of climate change is the availability of potable water for the consumption of growing population. The United Nations assessment has indicated that global water situation will become a major crisis in next fifty years in most of the countries of the world. Nearly 40% of the world's population is already under water stress condition. In spite of being a renewable resource, water is also a finite resource and vulnerable to depletion either due to pollution or loss into the sea without proper management.

Rivers and Wetland Support System

Northeast India receives intensive rainfall giving rise to several perennial rivers, streams, springs and lakes. South Assam is surrounded by hills on its three sides and receives natural water through several rivers like, Barak, Dhaleshwari, Jiri, Chiri, Katakhal, Gaghra, Madura, Rukni, etc. and there are several lakes and wetlands like Sonbeel, Chatla haor, Annua-Banskandi, Annua-Ramnagar, etc. and innumerable small ponds and streams. But much of its water resource remains unutilized due to lack of proper planning and systematic management. In spite of the abundance of water, there is scarcity of drinking water for the people and lack of irrigation facility for agriculture and tea plantation. It is indeed sad that the vast potential of water resource in the Barak valley region is yet to be harnessed for mini hydel-power projects, irrigation, aquaculture, fishery, aquatic diversity, hydroponics, aquatic bio-diversity, water tourism, etc. due to total lack of initiative for long term planning and water management policy leading to the drain off of water by rivers flowing to Bangladesh. On the other hand, the region has been experiencing frequent floods during rainy seasons and the draught-like situations during winter and spring months. Water stress in Barak valley has also been impounded by the wanton anthropogenic and biotic interferences.

The rivers and streams coming to the Barak valley region from the hills of Meghalaya, North Cachar Hills, Manipur, Mizoram and Tripura are experiencing large scale degradation due to deforestation and soil-depleting *jhum* cultivation. These human activities have denuded the hills and reduced the forest cover diminishing its water retention capacity which is responsible for siltation of rivers and lakes. There is hardly any planned programme to protect the catchments of the Barak river, which is affecting the quality and quantum of supply of water to wetlands in the valley. These wetlands once used to provide plenty of water for drinking, fishery, irrigation and fibre crops. In recent years, the fast growth of urbanisation, unauthorised encroachment of wetlands for agriculture by the migratory human population and negligence of the water bodies by the government have caused loss of invaluable bio-diversity and safe source of drinking water for the wild animals and local communities. Most of these wetlands and small water bodies have been polluted by organic and red algal blooms dominated by Euglena and other Cyano bacteria species, making the water unsuitable for drinking and domestic uses. Therefore, there is a need to conserve the wetlands and to develop proper schemes to protect these natural water bodies and ensure the supply of safe and plentiful water for drinking, agriculture, aquaculture, etc., rather than concentrating on the setting up of much talked about water-bottling plants in the region. The involvement of NGOs and appropriate official policy on the part of the government are the needs of the hour for harnessing and conservation of the natural freshwater bodies for the maintenance of the microclimatic condition of the region.

The situation calls for urgent strategies for protection of wetlands by involving inventory and integrated approach for classification of the catchments as critical water reservoirs. Water harvesting both on community basis and individual basis is needed for capturing the run off from rooftops, from local catchments, flood water and small streams and conserving those into the watersheds. The government may also work out public–private partnership platform to address the future need of water for survival of the people. Environment friendly farming system for sustenance of wetlands needs environmental awareness on the part of tea gardens, fishing communities, *haor*-dwellers and all categories of rural masses and involvement of the village panchayats, women self-help groups and other non-governmental agencies. The people have to be sensitised about the delicate relationship between human and environment and persuaded to participate in wetland conservation and management of water resources within the localities. Indeed, there is no alternative to community participation in conversation efforts.

Most of the wetlands in the Barak valley have natural channel linkages with the nearby rivers. These natural channels act as sustenance of life forms and rich bio-diversity. The riverine and wetland characteristic trees like *bonboa ceiba*, *anthocephabus codamba*, *eugenia jambolain*, and *lagerstomia flosreginae*, etc., along with cane and bamboo species, formed important timber resource and basis of many traditional household economic activities, besides maintaining the natural landscape. Over exploration of water resource has degraded the habitats and natural vegetations, resulting in river bank erosion and floods even during winter months. The wetlands and rivers have also been overexploited over years for fish catching by the *haor*-dwellers, adversely affecting the morphological and biotic characteristics of the wetlands. The situation has come to such a pass that eco-friendly sustainable measures are urgently required to minimise the river bank erosions and shrinking of the wetlands. The traditional technology for maintenance of the water bodies needs systematic encouragement and the conservation of aquatic bio-diversity in wetlands and *annuas* (oxbows) should also be given priority.

Strategies for Water Resource Management

Adequate knowledge, information and sound technology seem to be painfully wanting for understanding the tectonically unstable geo-hydrology of the rivers systems of Northeast India. As a result, the acceptability of large hydro projects is sometimes considered favourably by the authorities without scientifically examining the environmental feasibility of the projects. On the other hand, the integrated water resource management approach requires the involvement and participation of the stakeholders in the decision making processes. As per this approach, water resource management is to be accepted as a natural right of the people inhabiting the areas, both in upstream and downstream of the project sites. They should be fully informed about the consequences of the projects. Environmental feasibility is another requirement, which should be ascertained on the basis of the opinion of the experts who have adequately studied the areas that are likely to be impacted by the proposed projects. To facilitate this process, it is felt that there is a need to establish a Northeast Hydrometerorological Information System (NE-HIS) for data disseminations across the states of the region and national and international boundaries. Regional cooperation and coordination have to be established between the countries like China, Bhutan, Bangladesh and India for successful planning, management and utilization of water resources. There is also a great need for training and education for different target groups within the Northeast, both at formal and non-formal levels, to manage the water resources. This is necessary to overcome the problem of under-utilization of the available water, on one hand, and the sufferings of people due to frequent flood-damages to crops, on the other. Awareness of the people on all aspects of water conservation and environmental implications of the water bodies will be of immense help in achieving the cherished goal of sustainable development of water resources. The formation of a zonal watershed management authority and formulation of regional plans can go a long way in ensuring optimum utilization of water resources and minimising the recurrent natural disaster hazards, especially through strong infrastructural support systems. Long term integrated planning for river basin management is also required. Rain water harvesting is needed to rejuvenate the traditional water reservoirs. Human and natural environment should also be congenial for the maintenance of ecological balance. The zonal authority can be empowered to take care of all those parameters. Finally, only such projects which are environment friendly, which can fully involve the local communities, and which do not cause any inconvenience to the neighbouring states and countries should be considered for water resource related development in Northeast India.

REFERENCES

Ao, Temsula (2010), 'Land Ethics and Eco-Management Among the Ao-Nagas : With Special Reference to Chanki Village', *Northeast Researches,* Vol. I, Kolkata, 2010, pp. 7-16.

Baisya, D. (2010), 'Traditional Practices of Environmental Management in North East India : A Holistic way for prevention of Global Warming', In *climate change, global warming and N.E.India, Regional Perspective,* ERD Foundation, Guwahati, pp 34-43.

Brahmaputra Board (2000), *Water vision. North East 2050. Report of the Brahmaputa Board,* Ministry of Water Resources, Govt. of India, 78 pp.

Das, P.J. (2004)., 'Rainfall regime of Northeast India : A hydrometeorological study with especial emphasis on the Brahmaputra basin', Ph.D thesis, Gauhati University, Guwahati.

Das P.J. and Goswami, D.C. (2003), 'Long Term variability of rainfall over northeast India', *Indian Journal of landscape systems and Ecological Studies* 26(1), 1-20.

Sarma, H.P. (2010), 'Climate change and Water Resources of North East India', *Climate Change, Global Warming and NE India, Regional perspective,* ERD Foundation, Guwahati, pp 12-16.

Verghese, B.G. (2006)., 'Water Resources in the North East : Development options in a cooperative Framework'. WWW. North-East of India (PDF).

4

An Overview of the Ecology of Some Streams and Rivers of Northeast India

Tapati Das & Abhik Gupta

[**Abstract:** The paper reviews the information available on the ecology of some selected streams and rivers in three North Eastern states of India within a span of 25 years. These are rivers Umkhrah, Umshirpi and Umkaliar in Meghalaya, river Barak in Assam and rivers Pare, Pachin and Dikrong in Arunachal Pradesh. All these rivers are important sources of usable water by the local people in respective areas and are also the abodes of diverse varieties of aquatic organisms including fishes. Some of the common disturbances that are observed along the river basins of Northeast India include siltation, unplanned urbanisation and the associated input of sewage and municipal wastes into the rivers, besides picnicking, quarrying activities, open defecation on the river bank, disposal of industrial effluents and so on. Nevertheless, all these anthropogenic disturbances are likely to result in the impairment of water quality, habitat destruction and loss of biodiversity in the riverine systems in the region which also holds the status of one of the biodiversity hotspots in the world. Overall, the comparisons of physico-chemical and biological parameters of these lotic systems reveal the presence of organic pollution rather than marked industrial interferences. The main reason which can be ascribed for such incidences might be due to unplanned urbanization, human disturbance, and ignorance of the local people regarding ecological services the rivers provide. Therefore, the situation warrants the need to develop a long-term policy for the management of the river systems by integrating ecological and social sciences with landscape design.]

Introduction

Northeast India falls within the tropical monsoon climate with an average annual rainfall of 200 cm. This region is also endowed with the highest rainfall area in the world, located in Mawsynram-Cherrapunjee of Meghalaya. Besides, Northeast India has one of the densest drainage networks in the world because of high rainfall accompanied by the character of the topography with hills, mountains and flat lands. In Northeast India there are a total of seven drainage systems. They are (i) the Brahmaputra in the north, (ii) Barak-Surma system in the middle, (iii) Chindwin system fed by the Tezu, Yu and Manipur rivers in the east and southeast, (iv) Chhimtuipui in the southern part, (v) Karnaphuli in western part of Mizoram, (vi) Gumti in the middle part and (vii) Feni in the southern part of Tripura. All these drainage basins ultimately find their way either into Bangladesh or Myanmar. Therefore, the drainage basins of Northeast India are important also from international standpoint. Nevertheless, amongst all the drainage basins Brahmaputra and Barak are very much extensive and are the major drainage basins in this region.

Some of the common features that are observed along the river basins of Northeast India in relation to disturbance are siltation, unplanned urbanisation and the associated input of sewage and municipal wastes into the rivers, besides picnicking, quarrying activities, open defecation on the river bank, disposal of industrial effluents and so on with certain minor differences in the type of anthropogenic interferences amongst the states. Nevertheless, all these anthropogenic disturbances are likely to result in the impairment of water quality, habitat destruction and loss of biodiversity in the riverine systems in the region which also holds the status of one of the biodiversity hotspot regions in the world.

In the present paper, an attempt has been made to compare the ecology and aquatic diversity of some selected rivers in the three North-Eastern states of India within a 25 years time span

Table 1. Anthropogenic disturbances in some selected riverine systems of Northeast India

Disturbances	*Meghalaya*	*Assam*	*Arunachal Pradesh*
Landslides, land slips and siltation	+	+	+
Settlements on the river bank	+	+	+
Grazing along the riparian region	+	+	+
Disposal of piggery wastes on the river bank	+	-	+
Defecation along the river bank	+	+	+
Agriculture along the river bank	+	+	+
Discharge of sewage and storm drainage into the river	+	+	+
Washing cloths and human bathing	+	+	+
Cattle washing	-	+	+
Washing vehicles	-	-	+
Quarrying of boulders, cobbles and pebbles	+	-	+
Quarrying of sand	+	-	-
Coal mining activities on the river bank	+	-	-
Running of vehicles along the riparian region	+	+	+
Urbanization along the watershed	+	+	+
Market complex along the watershed	+	+	+
Discharge of industrial effluents	-	+	-

Source: Gupta, 2006; Das *et al.*, 2007; Das and Gupta, 2008.

or so. However, a total comparison is not possible as neither the sites nor the parameters examined tally completely. The present observation stands on the results based on the investigations of Gupta and Michael (1983), Ao *et al.* (1984), Gupta (1994) and Gupta (2006) in some rivers of Shillong, Meghalaya, Das (2001), Das and Gupta (2008) in Barak Valley, Assam, and Das *et al.* (2007), and Bagra *et al.* (2009) in Arunachal Pradesh.

In the present case seven rivers and streams were selected for their limnological investigation, viz., R. Umkhrah, Umshirpi and Umkaliar in Meghalaya, R. Barak in Barak Valley, Assam and R. Pare, Pachin and Dikrong in Arunachal Pradesh. All these rivers are important sources of usable water by the local people in the respective areas. It may be mentioned here that while selected rivers in Meghalaya and Arunachal Pradesh belong to the Brahmaputra river basin, the rivers in Barak Valley belong to the Barak river basin.

Disturbances in the Study Area

The disturbance regime in the study area (Table 1) shows that all the rivers in the study area encountered nearly similar type of disturbances, like siltation, unplanned urbanisation and open defecation along the river bank, with only few exceptions. This might be due to the cultural differences amongst the local people as well as the availability of riverine resources like boulders and sand in the study area.

Physico-chemical properties of water in the riverine systems

Physico-chemical properties of water in the study area (Table 2) reveal variations in the river water properties in the three states. When certain water variables were compared with relevant Indian standards for drinking water after disinfection (as per IS:2296), it was observed that some streams of Meghalaya showed unsatisfactory status in terms of water pH, BOD and presence of heavy metals like cadmium, copper, manganese, lead and zinc. On the other hand, water quality in river Barak was within the prescribed limit at least for the parameters which were taken into consideration during the study period. And, rivers of Arunachal Pradesh revealed greater BOD than the threshold limit for the drinking water quality. These results therefore, revealed the presence of greater input of organic matters into the rivers. This might be due to unplanned urbanization, input of sewage and municipal wastes either directly or entering into the riverine systems along with runoff water.

Table 2. Physico-chemical properties of water in some selected riverine systems of Northeast India

Parameters	*Meghalaya*	*Assam*	*Arunachal Pradesh*	*Indian Standard (as per IS:2296)*
Air temperature (°C)	13.4-23.7	29.08-32.21	25.88-28.81	-
Water temperature (°C)	7-24	25.38-27.04	18.60-24.14	-
Transparency (cm)	-	20.33-27.94	Transparent-21.45	-
Water flow rate (m sec^{-1})	0.22-0.85	-	0.94-1.37	-
Total dissolved solids (mg l^{-1})	34-265	33.73-45.09	46.75-102.13	500
pH	6.1-7.3	7.19-7.74	7.03-7.32	6.5-8.5
Conductivity (µS cm^{-1})	12.7-183	67.45-90.18	60-120	-
Total alkalinity (mg l^{-1})	10-100.8	54.37-69.03	58.16-68.69	-
Dissolved oxygen (mg l^{-1})	6-8.7	7.21-7.97	8.24-10.84	6 (min)
Biological oxygen demand (5 days at 20°C) (mg l^{-1})	ND-45.7	-	0.88-3.13	2
Chemical oxygen demand (mg l^{-1})	ND-162	-	-	-
Free carbon dioxide (mg l^{-1})	-	1.29-1.98	2.88-5.46	-
Chloride (mg l^{-1})	9-31.5	-	1.85-12.42	250
Nitrate-N (mg l^{-1})	Trace-8	1.36-2.16	0.14-0.29	20
Ammonium –N (mg l^{-1})	-	-	0.07-0.24	-
Phosphate-P (mg l^{-1})	-	-	0.02-0.04	-
Sodium (mg l^{-1})	-	6.13-8.56	-	-
Potassium (mg l^{-1})	-	2.16-6.64	-	-
Cadmium (mg l^{-1})	<0.5-4.5	-	-	0.01
Copper (mg l^{-1})	<1.0-25.5	-	-	1.5
Manganese(mg l^{-1})	<1.0-204.5	-	-	0.5
Lead (mg l^{-1})	0-27.3	-	-	0.1
Zinc (mg l^{-1})	4.3-37.6	-	-	-

Source: Gupta and Michael, 1983; Gupta, 2006; Das *et al.*, 2007; Das and Gupta, 2008.

Table 3. Distribution of phytoplankton in some selected riverine systems of Northeast India

Class	*Genus*	*Meghalaya*	*Assam*	*Arunachal Pradesh*
Bacillariophyceae	*Navicula*	+	+	+
	Achnanthes	-	-	+
	Cocconeis	-	-	+
	Coloneis	-	-	+
	Cymbella	+	-	+
	Fragillaria	+	-	+
	Mastogloea	-	-	+
	Meridion	-	-	+
	Nitzschia	+	-	+
	Pinnularia	+	-	+
	Synedra	+	-	+
	Amphora	+	-	-
	Frustulia	+	-	-
	Gomphonema	+	-	-
	Cyctella	+	-	-
	Melosira	+	-	-
	Hantzschia	+	-	-
Chlorophyceae	*Actinastrum*	-	+	+
	Microspora	-	+	+
	Spirogyra	+	+	+
	Closterium	+	+	+
	Chaetophora	-	-	+
	Chlorococcum	-	-	+
	Cladophora	-	-	+
	Cosmarium	-	-	+
	Cylindrocapsa	-	-	+
	Golankinia	-	-	+
	Hormidium	-	-	+
	Ophila	-	-	+
	Penium	-	-	+
	Rhizoclonium	-	-	+

	Sphaerozosma	-	-	+
	Trebanaria	-	-	+
	Ulothrix	-	-	+
	Volvox	-	-	+
	Staurastrum	+	-	-
	Netrium	+	-	-
	Pleurotaenium	+	-	-
	Scenedesmus	+	-	-
	Sphaeroplea	+	-	-
	Eudorina	+	-	-
	Phormidium	+	-	-
	Crucigenia	+	-	-
Cyanophyceae	*Oscillatoria*	+	+	+
	Anabuena	+	+	+
	Merismopedia	+	-	-
	Aphanocapsa	-	-	+
	Chroococcus	-	-	+
	Gloeothece	-	-	+
	Lyngbya	-	-	+
	Microcystis	-	-	+
	Nostoc	-	-	+
	Spirulina	-	-	+
	Synechocystis	-	-	+
Chrysophyceae	*Synura*	+	-	-
	Chlorobotrys	+	-	-
Dinophyceae	*Gymnodinium*	+	-	-
Rhodophyceae	*Bratrachospermum*	+	-	-
Euglenophyceae	*Euglena*	+	-	+
	Phacus	-	-	+
Xanthophyceae	*Gloeochloris*	-	-	+
	Chlorobotrys	-	-	+
Total: 8 classes	63 genera	30	7	43

Source: Ao *et al.*, 1984; Das, 2001; Das *et al.*, 2007.

Table 4. Distribution of zooplankton in some selected riverine systems of Northeast India

Class	*Genus*	*Meghalaya*	*Assam*	*Arunachal Pradesh*
Protozoa	*Paramaecium*	-	+	+
	Spirostomum	-	+	-
Rotifera	*Monostyla*	-	+	+
	Diplois	-	+	-
	Polyartha	-	+	-
	Lecane	-	+	+
	Lepadella	-	+	-
	Brachionus	-	+	+
	Testudinella	-	+	-
	Mytilina	-	-	+
	Philodina	-	-	+
Branchiopoda	*Moina*	-	+	+
	Simocephalous	-	+	+
	Ceriodaphnia	-	-	+
	Leydigia	-	-	+
Copepoda	*Mesocyclops*	-	+	-
	Neodiaptomus	-	+	-
	Cyclops	-	-	+
	Ectocyclops	-	-	+
Arachnida	*Hydracarina*	-	-	+
Total: 5 classes	20 genera	-	13	13

Source: Das, 2001; Das *et al.*, 2007.

Diversity Status in the Riverine Systems

Distribution of phytoplankton in the selected riverine systems (Table 3) depicts the presence of a total of 63 genera of phytoplankton with Chlorophyceae as the most dominant group. This indicates relatively fast flowing conditions accompanied by a low temperature in the region. Zooplankton revealed the presence of a total of 20 genera (Table 4) with Rotifera as the most dominating group, thereby indicating the occurrence of organic enrichment in those systems. Benthic macro invertebrates in the region represented a total of 71 families (Table 5) with Diptera as the most dominant order which reflects the greater incidence of the input of organic matter into the riverine systems. The distribution of fish in the region revealed the presence of a total of 67 taxa with Cyprinidae as the most dominant family. This is perhaps due to the greater tolerance of this group to a wide range of environmental conditions prevalent in the riverine systems.

Table 5. Distribution of benthic macroinvertebrates in some selected riverine systems of Northeast India

Class	*Order*	*Family*	*Meghalaya*	*Assam*	*Arunachal Pradesh*
Insecta	Ephemeroptera	Heptageniidae	+	-	+
		Leptophlebiidae	+	-	+
		Epemerellidae	+	-	+
		Caenidae	+	-	-
		Baetidae	+	+	+
		Ecdyonuridae	-	-	+
		Ephemeridae	-	-	+
		Potamanthidae	-	-	+
		Siphlonuridae	-	-	+
	Placoptera	Peltoperlidae	+	-	-
		Perlidae	-	-	+
		Perlodideae	-	-	+
		Pteronarcyidae	-	-	+
		Taeniopterygidae	-	-	+
	Trichoptera	Leptoceridae	+	-	-
		Brachycentridae	+	-	-
		Psychomyiidae	+	-	-

	Hydrophychiidae	+	+	+
	Molannidae	-	-	+
	Philopotamidae	-	-	+
	Rhyacophilidae 1	-	-	+
	Rhyacophilidae 2	-	-	+
Odonata	Agriidae	+	-	-
	Cordulegasteridae	+	+	+
	Aeshnidae	+	-	+
	Libellulidae	+	-	-
	Coenagriidae	+	-	-
	Gomphidae	-	-	+
Hemiptera	Mesovelidae	+	-	+
	Gerridae	+	-	-
	Notonectidae	+	-	-
	Corixidae	+	+	+
	Belostomatidae	-	-	+
	Nepidae	-	-	+
Diptera	Tipulidae	+	-	+
	Simuliidae	+	-	+
	Chironomidae	+	+	+
	Chironomidae (Chironomous)	-	-	+
	Culicidae	-	-	+
	Diptera pupa 1	-	-	+
	Diptera pupa 2	-	-	+
	Psychodidae	-	-	+
	Rhagionidae	-	-	+
	Stratiomyidae	-	-	+
	Diptera 1	-	-	+
	Diptera 2	-	-	+
Coleoptera	Dysticidae	+	-	-
	Gyrinidae	+	-	-
	Curculionidae	-	-	+

Table 5. (Contd.)

Class	*Order*	*Family*	*Meghalaya*	*Assam*	*Arunachal Pradesh*
		Beetle larva 1	-	-	+
		Beetle larva 2	-	-	+
		Coleopteran larva	-	-	+
		Psephenidae larvae 1	-	-	+
		Psephenidae larvae 2	-	-	+
		Psephenidae (adult)	-	-	+
		Semi aquatic beetle	-	-	+
		Coleoptera 1	-	-	+
	Megaloptera	Corydalidae	+	-	+
		Placopteris	-	-	+
Crustacea	Decapoda	Palimonidae	-	+	+
		Astacidae	+	-	-
Gastropoda	Basommatophora	Lymnaeidae	+	-	-
		Planorbidae	+	-	-
	Neotaenioglossa	*Thiaridae*	-	+	+
	Architaenioglossa	Viviparidae (*Angulyagra oxytropis*)	-	+	+
Oligochaeta		Archioligochaeta Tubifex	+	+	-
	Neo-oligochaeta	Lumbriculidae	-	-	+
Clitellata	Hirudinea	Hirudidae	+	-	-
Nematoda	Rhabditoidea	Rhabditidae	-	+	+
Arachnida	Trombidformes	Arrenuridae (*Hydrocarina* sp.)	-	+	-
Turbellaria	Seriata	Planariidae	+	-	+
Total: 8 classes		18 orders 71 families	31	11	53

Source: Gupta, 1994; Das, 2001; Das *et al*., 2007.

Table 6. Distribution of fish in some selected riverine systems of Northeast India

Order	*Family*	*Taxa*	*Meghalaya*	*Assam*	*Arunachal Pradesh*
Beloniformes		Belonidae *Xenenthodon cancila*	-	+	-
Clupeiformes		Clupeidae *Gudusia chapra*	-	+	+
		Hilsa ilisha	-	+	-
	Strigeidae	*Trichogaster fasciatus*	-	+	-
Cypriniformes		Cyprinidae *Aspidoparia jaya*	-	-	+
		Bangana dero	-	-	+
		Barilius bendelisis	-	-	+
		Cirrhinus reba	-	+	+
		Danio rerio	-	-	+
		Esomus danricus	-	-	+
		Puntius sophore	-	-	+
		Salmophasia bacaila	-	-	+
		Schizothorax progastus	-	-	+
		Semiplotus semiplotus	-	-	+
		Labeo pangusia	-	-	+
		Labeo rohita	-	+	+
		Labeo calbasu	-	+	-
		Labeo gonius	-	+	-
		Garra gotyla	-	-	+
		Amblypharyngodon mola	-	+	-
		Catla catla	-	+	-
		Cirrhinus mrigala	-	+	-

(contd.)

		Puntius sp.	-	+	-
		Tor sp.	-	+	-
	Psilorhynchidae	*Psilorhynchus homaloptera*	-	-	+
		Psilorhynchus arunachalensis	-	-	+
		Schistura sp	-	-	+
		Physoschistura sp	-	-	+
		Botia Dario	-	-	+
		Canthophrys gongota	-	-	+
		Lepidocephalus guntea	-	-	+
Osteogloss iformes	Notopteridae	*Notopterus chitala*	-	+	-
		Notopterus notopterus	-	+	-
Perciformes	Ambassidae				
		Chanda nama	-	+	-
	Anabantidae	*Anabas testudineus*	-	-	+
	Gobiidae	*Glossogobius givrius*	-	+	-
Siluriformes	Amblycipitidae	*Amblyceps* sp	-	-	+
	Bagridae	*Aorichthys aor*	-	+	-
		Mystus cavasius	-	+	+
		Mystus bleekeri	-	-	+
		Mystus vittatus	-	+	+
		Mystus montanus	-	-	+
		Mystus sp	-	-	+
		Rita rita	-	+	-
	Chacidae	*Chaca chaca*	-	+	+
	Clariidae	*Clarias batrachus*	-	-	+
	Erethistidae	*Erethistes pussilus*	-	-	+

(Contd.)

		Hara hara	-	-	+
		Hara jerdoni	-	-	+
		Hara sp	-	-	+
	Heteropneustidae	*Heteropneustes fossilis*	-	-	+
	Olyridae	*Olyra longicaudata*	-	-	+
	Pangasiidae	*Pangasius pangasius*	-	+	-
	Schilbeidae	*Clupisoma garua*	-	+	-
		Eutropichthys vaccha	-	+	-
		Ailia coila	-	+	-
	Siluridae	*Ompok pabo*	-	-	+
		Pterocryptis gangelica	-	-	+
		Ompok sp.	-	+	-
		Wallagu attu	-	+	-
	Sisoridae	*Gogangra viridescens*	-	-	+
		Nangra assamensis	-	-	+
		Bagarius bagarius	-	+	-
Synbran chiformes	Symbranchidae *cuchia*	*Monopterus* -	-	+	
		Badis sp	-	-	+
	Mastacembelidae	*Mastacembelus armatus*	-	+	-
Tetraodo- ntoformes	Tetraodontidae *cutcutia*	*Tetraodon*	-	-	+
Total: 8 orders	23 families	67 taxa	-	30	43

Source: Das, 2001; Bagra *et al.*, 2009.

Conclusions and Recommendations

Overall, the present comparisons of physico-chemical and biological parameters of some major rivers of three states of Northeast India reveal the presence of organic pollution rather than the industrial interference. The main reasons which can be ascribed for such incidence include unplanned urbanization, poverty and ignorance of the local people regarding the ecological services the rivers provide. Therefore, this study warrants the need to develop a long-term strategy for the management of the riverine systems by integrating ecological and social sciences with landscape design. This can be made possible only by involving local people from stakeholders to policy makers in the river management.

Some of the management activities of the riverine systems in this part may be in terms of:

(i) Restriction of the dumping of garbage and other organic wastes into the rivers.

(ii) Scientific planning of at least the remaining township area in the region.

(iii) Installation of proper sewage treatment plants.

(iv) Afforestration on the river bank and riparian zone so that there is control to some extent of the siltation into the riverine systems.

(v) Providing the poor with low cost sanitation facilities.

(vi) Acquainting the local people about ecosystem services that the rivers provide and at the same time the poor should be provided with alternate sources of livelihood so that the pressure on the riverine system is minimised.

(vii) Some alternative sources of building materials for construction purposes so that excess pressure on the riverine systems in terms of quarrying activities is relieved. In order to release pressure on the river bank and riparian region there should be improvement of grazing lands by the introduction of grasses and legumes which otherwise causes soil erosion and ultimately siltation due to the trampling effects of livestock.

(viii) All the medium and small scale industries should take proper environmental measures so that the disposal of the effluents is after its proper treatment.

(ix) And, last but not the least, there should be implementation of the principle 'polluter pays'.

REFERENCES

Ao. M., Alfred, J. R. B. and Gupta, A. (1984), 'Studies on some lotic systems in the North-eastern Hill regions of India', *Limlogica* (Berlin) 15(1): 135-141.

Bagra, K., Kadu, K., Sharma, K. N., Laskar, B. A., Sarkar, U. K., Das, D. N. (2009), 'Lists of species, Ichthyological survey and review of the checklist of fish fauna of Arunachal Pradesh, India', *Check List* 5(2): 330–350.

Das, T. (2001), 'A study on the limnology and capture fisheries of river Barak, Assam', M. Phil. Dissertation, Assam University, Silchar, 151 pp.

Das, T., Arunachalam, A., Gupta, G. and Arunachalam, K. (2007), 'Biodiversity of plankton and macroinvertebrates in Dikrong river system in the Indian Eastern Himalaya', In: *Biodiversity Utilization and Conservation* (eds. Arunachalam, A. and Arunachalam, K., Aavishkar Publishers Distributors, Jaipur, pp. 220-236.

Das, T. and Gupta, A (2008), 'An appraisal on the physico-chemical properties of water along a disturbance gradient in river Barak, Assam, north east India" *The Bioscan* 3(1): 109-114.

Gupta, A. and Michael, R. G. (1983), 'Seasonal differences and relative abundance among population of benthic aquatic insects in a moderately high altitude stream'. *Proc. Wkshp. High. Alt. & Wildl. Ecol. Zool. Surv. India*, pp. 21-28.

Gupta, A. (1994), 'A comparative assessment of methods for biological surveillance of running water quality using macroinvertebrates', In: *Proc. 3rd National Symposium on Environment, Thiruvanthapuram*, (eds.), Nair, N. B., Eapen, C. D., Bapat, V. N., Sadasivan, S. and Gangadharan, P., Board of Research in Nuclear Sciences, Dept. of Atomic Energy, Govt of India, pp. 209-213.

Gupta, A. (2006), 'Streams of Shillong. An Ecological Overview', In: *Rivers and Riverine Landscape in North East India*' (eds.), Sengupta, S., Sinha, K. R., Das, S., Rani, W. B. and Purkayastha, S., Concept Publishing Company, New Delhi, pp. 118-127.

5

Management of Water Resources of Barak Valley for Sustainable Economic Development

Niranjan Roy & Suparna Roy

[**Abstract:** The backwardness of the Barak Valley region in manufacturing activities can be gleaned from the occupational distribution of population, more than 70 percent of which is dependent on agricultural sector. The river Barak with a catchments area of 78.150 sq.km traverses a total distance of 900 km, out of which 532 km lies in India and the rest flows in Bangladesh. There is no flood problem in the first reach of 403 km., from source to Lakhipur, but from Lakhipur to Bhanga (129 km] there is acute problem of flood and drainage congestion. Flood is a recurrent problem in the valley. Every year large areas come under the grip of floods, causing extensive damages to properties and crops and disrupting the communication system. On the other hand, the irrigation facility is practically non-existent and whatever meager facilities are available, the utilization ratio is very low. It has been found from the sample study that the productivity in irrigated farms is much higher compared to non-irrigated farms. A suitable policy for expansion of irrigation on one hand and control of flood on the other should be undertaken by the government at an early date. The emphasis should be to develop some medium and large irrigation projects to be managed entirely by the farmers. Since the Tipaimukh River Valley Project seems to have been abandoned due to pressure from Bangladesh, an alternative flood control measure should be undertaken at the earliest.]

Introduction

The Barak Valley, which forms the southern geographical division of the state of Assam in Northeast India, is surrounded by hills on three sides and has a riverine international border with Bangladesh on the other side. Before partition and independence of India, it was a part of the larger Surma Valley famed for its tea industry under the British colonial rule. But at the time of independence the Sylhet district of Surma Valley was transferred to East Bengal to form East Pakistan, which was to become Bangladesh later on. As a result, the area now popularly identified as Barak Valley (consisting of Cachar, Hailakandi and Karimganj districts) lost its vital road and waterway links to the rest of India through East Bengal. Though the rail and road connections with the Brahmaputra Valley in the north are in existence since the British time, the surface communication of the region with the outside world through the mountainous routes (via Khasi-Jaintia Hills and the North Cachar Hills) has always been difficult and prone to intermittent disruptions. The traditional tea industry has been on decline since independence and the initiatives for setting up alternative new industrial units have largely proved unsuccessful. Backwardness of the region in manufacturing activities can be understood from the occupational distribution of population in the valley. More than 70 percent of the population is dependent on agricultural sector. Therefore, sincere efforts should be made for the sustainable economic development of the region based on local resources, besides revival of the traditional tea sector.

Economy of Barak Valley

In 2000-01, Barak Valley accommodated 11.71% of the total population of Assam while its share in the Gross State Domestic Product (GSDP) was only 9.33%. The district per capita income is also below the state average, which in turn is much below the national average. As per statistics of 2000-01, the per capita income (i.e., net district domestic per capita) in Barak Valley was Rs. 5223 at 1993-94 prices. The figure comes up to only about 84% of the corresponding per capita income figure for Assam. The sectoral composition of the economy of the valley confirms strong agricultural bias. Agriculture contributed 33.1% Gross District Domestic Product (GDDP) and the primary sector contributed 38.3% of the same. The tertiary sector comprising of the various services constitutes more than one third of the GDDP. The manufacturing sector's share is less than 8.7% of the secondary sector as a whole and it accounts for 9.49% of the GDDP. The relatively high share of secondary sector (higher than that of the manufacturing sector) is due to a relatively higher contribution of construction activities. The sectoral composition of GDDP of Barak Valley has a strong resemblance to the sectoral composition of Gross State Domestic Products (GSDP) of Assam. For instance, for the state of Assam as a whole agriculture and primary sector still constitutes the mainstay. Moreover, in the secondary sector the construction activities played an important role.

The relevant facts clearly show that agriculture is the principal occupation of the people of the Barak Valley. About 70% of the people are directly dependent on agriculture. On the other hand, a striking feature of the cropping pattern of the Barak Valley region is the predominance of paddy, which accounted for more than 90% of the total cropped area during the period 1971-72 to 2004-05. The share of paddy during these periods fluctuated between 91.64% and 95.12 % without showing any distinct trend. Other notable crops cultivated in the valley are pulses, rape and mustard seeds, potato and sugarcane (Roy and Bezbaruah, 2002).

The comparative picture of average yield of rice in the Barak Valley region, the state of Assam as a whole and at the all-India level emerges from Table-1. It shows some changes between the years within the period 1977-78 to 2000-01. However, the broad fact that emerges is that the Barak Valley yield has been above the Assam yield for most of the years, though the former in general remained below the all-India average yield. The Barak Valley yield as a proportion of the all-India yield also shows wide fluctuations ranging between 55% and 112%. This large instability may be attributed to the fact that in Barak Valley, paddy is still grown almost entirely under rain-fed conditions and the irrigation infrastructure is practically non-existent in the region. This instability notwithstanding, the gap of yields of rice in the Barak Valley region and at the all-India level did not show any distinct tendency to widen over the reference period of our study.

Table 1: Comparative Average Yield of Rice Crop in Barak Valley, Assam and All India during the period 1977-78 to 2004-05

Year	*Rice Yield (kg/hectare)*			*Barak Valley yield as percentage of all India yield*	*Barak Valley yield as percentage of all Assam yield*
	Barak Valley	*Assam*	*All India*		
(1)	*(2)*	*(3)*	*(4)*	*(5)*	*(6)*
1977-78	1238	1034	1308	94.65	119.73
1980-81	1497	1120	1336	112.05	133.66
1981-82	910	1006	1308	69.57	90.46
1982-83	1276	1130	1231	103.66	112.92
1983-84	1236	1145	1457	84.83	107.95
1984-85	1107	1064	1417	78.12	104.04
1985-86	1241	1173	1552	79.96	105.80
1986-87	1184	1056	1471	80.49	112.12
1987-88	1556	1180	1465	106.21	131.86
1988-89	1471	1095	1689	87.09	134.34
1989-90	963	1135	1745	55.19	84.85
1990-91	1387	1292	1740	79.71	107 35
1991-92	1634	1290	1751	93.32	126.67
1992-93	1661	1341	1744	95.24	123.86
1993-94	1291	1331	1888	68.38	96.99
1994-95	1677	1350	1921	87.30	124.22
1997-98	1650	1359	1900	86.84	121.41
1998-99	1721	1345	1921	89.59	127.96
2001-02	1730	1427	1913	90.43	121.23
2004-05	1534	1513	1882	81.15	101.38

Sources: 1) Estimates of Area, Production and Average Yield of Principal Crops in Assam from 1951-52 to 1987-88, Directorate of Economics and Statistics, Govt. of Assam.

2) Basic Agricultural Statistics, Directorate of Agriculture, Govt. of Assam, various issues.

3) Agricultural Statistics at a Glance, Directorate of Economics and Statistics, Department of Agriculture Co-operation, Ministry of Agriculture, Govt. of India, 2005.

River Systems

The principal river of the Barak Valley region, namely, the Barak, is the second largest river in the North-Eastern Region (NER) of India. It is a major tributary of Meghna River of the great Ganga-Meghna-Brahmaputra system. The river took its rise a little to the west of Maothana, on the northern boundary of Manipur. Taking a south-west turn from its source near Tipaimukh in Manipur, the river then comes to north and for a considerable distance forms the boundary line between the Cachar district of Assam and the state of Manipur. The river Jiri, one of the major tributaries of Barak, joins the Barak in this sector and even after the junction the combined stream for a considerable area forms the border between Cachar and Manipur. At Jiribam, it turns again towards the west and flowing through the heart of Cachar district (and touching the Silchar city) reaches Badarpur town in the Karimganj district. From Badarpur to Haritikar, it provides the boundary between the Karimganj district and Bangladesh. At Haritikar, the river divides into two branches, namely, Surma and Kushiara. The later branch immediately enters Bangladesh while the former passes through the Karimganj town and continues to form the boundary between India (Barak Valley) and Bangladesh (Bhattacharjee, 1977).

The river Barak, with catchments area of 78.150 sq. km., traverses a total distance of 900 km, out of which 532 km length lies within India and rest in Bangladesh. It flows in a deep channel through a narrow valley, varying 25-30 km wide. The different reaches of Barak and its morphological characteristics are shown in Table-2. The river enters the Cachar plain near Lakhipur. Its northern tributaries are Jiri, Chiri, Madhura, Jatinga, Harang and Kalain, and the major southern tributaries are Sonai, Katakhal and Dhaleswari. The other principal rivers of the region are Rukni (tributary of Sonai), Longai, Singla and Kakra. Most of these rivers dry up during winters and due to deposit of silts the drainage capacity becomes poor in all streams. The rivers are hardly navigable even in the rainy season. In summer, all these rivers rise up in high spate causing heavy floods almost every year. There is no flood problem in the first reach of 403 km. from source to Lakhipur, since the main river and the tributaries have

Table 2: Morphological Characteristics of River Barak

Reach	Reach Length	Slope	Channel characteristics
(i) upper reach from source of Tipaimukh	312 km	1 in 70	Very deep channel and narrow
(ii) Reach from Tipaimukh to Lakhipur	91 km	Varies from 1 in 3000 to 1 in 21,500	Flows in hilly terrain
(iii) Lakhipur to Indo-Bangladesh Border	129 km (Average)	Flatter Slope 1 in 21,500	Alluvial plain meandering
Total	532 km		

Source: Water Resource Department, Govt. of Assam, 2004.

high banks in that hilly reach. The adverse topography and heavy rainfall have made the valley region liable to frequent floods. The baneful capacity of the river Barak is about 4260 cumecs while the maximum discharge observed so far is 67000 cumecs (Bezbaruah, 1997).

Flood Affected (FA) and Non-Flood Affected (NFA) Areas

For estimating the FA and NFA areas in the Revenue Circles the names of Flood Affected (FA) and Non-Flood Affected (NFA) villages have been collected from the official records available with the Circle Officers of the Revenue Circles. The details of the Flood Affected and Non-Flood Affected villages of the valley are shown in Table 3. The analyses are based on the census data 2001 and the data collected from all the Revenue Circles of the three districts of the valley. There are 2355 villages in the three districts of the Barak Valley. Out of a total of 2355 of villages in the three districts of the valley, 876 i.e. 37.2% of the villages are flood affected but 43.8% of the total population and 43% of total households of the valley are flood affected. Among the three districts in the valley, Karimganj possesses more flood affected areas than the other two districts. Out of a total of 876 villages, the Karimganj district alone accommodates 447 i.e 51.03% of total flood affected villages of the valley. While Cachar distrct accommodates 30.71%, Hailakandi district possesses 18.26% of total flood affected villages. In respect of Area, Households and Population also Karimganj district accommodates more flood affected inhabitants than the other two districts (Roy, 2009).

Flood Damage

Flood is a recurrent problem in the Barak Valley. Every year large areas come under the grip of floods, which cause extensive damage to properties and crops and disrupt the communication system. Flood and water stagnation may occur any time from May to October, the chief reasons being high rainfall, high catchments area due to surrounding hills, silting up of river beds, etc. The crux of the problem is that once flood water or rainwater deposition occurs, the stagnant condition prevails for months together in most places due mainly to lack of effective drainage system in the region. The principal river of the region, Barak, and its branches, Surma and Kushiara, receiving discharges from the catchments areas and spill over water from them as well as back flow of water to their respective tributaries are responsible for the flood situation.

Table-3 shows the area affected and crops area damaged by floods in the Barak Valley region during the period 1982 to 2004. The figures in the table clearly show that barring only three years during the period of our study flood visited every year in the region. In many years of the reference period there were three or even four waves of flood. So far as damage to crop area is concerned, the highest flood during the recent time was experienced in 1989 in which 42.90% of crop area to total estimated gross cropped area and about 23.51% of reporting area were affected by floods, while during the year 1988, 26.47% of crop area and 32.12% of reporting area were damaged. One important fact that emerges from the table is that there is no direct correspondence between the extent of area affected by flood and the extent of damage of the crops. While in some years crop damage is comparatively

Table 3: Area, Household and Population in FA and NFA Areas of Three Districts of Barak Valley

District	*Number of villages*			*Area (in Hectare)*			*Household*			*Population*		
	FA	*NFA*	*Total*	*FA*	*NFA*	*Total*	*FA*	*NFA*	*Total*	*FA*	*NFA*	*Total*
Cachar	269 (11.40)	786 (33.40)	1055 (44.80)	59404 (12.70)	152829 (32.70)	212233 (45.40)	76101 (15.50)	158225 (32.30)	234326 (47.80)	409304 (15.20)	834230 (31.10)	1243534 (46.30)
Karimganj	447 (19.00)	522 (22.20)	969 (41.20)	61677 (13.20)	119938 (25.60)	181615 (38.80)	86441 (17.60)	80353 (16.40)	166794 (34.00)	492403 (18.30)	449685 (16.80)	942088 (35.10)
Hailakandi	160 (6.80)	171 (7.20)	331 (14.10)	40615 (8.70)	33482 (7.10)	74097 (15.80)	48508 (9.90)	40688 (8.30)	89196 (18.20)	275428 (10.30)	223359 (8.30)	498787 (18.60)
Barak Valley	876 (37.20)	1479 (62.80)	2355 (100.00)	161696 (34.60)	306249 (65.40)	467945 (100.00)	211050 (43.00)	279266 (57.00)	490316 (100.00)	1177135 (43.80)	1507274 (56.20)	2684409 (100.00)

Source: Primary data collected by the authors, 2009

Table 4

Area Affected and Crop Area Damaged by Floods in the Barak Valley Region during the period 1982 to 2004

Year	*Area affected*		*Crops area affected*	
	in hectares	*as percentage of reporting area*	*in hectares*	*as percentage of estimated Gross Cropped Area*
(1)	(2)	(3)	(4)	(5)
1982	2,677	0.39	1,137	0.45
1983	1,37,384	19.88	22,533	8.70
1985	3,72,500	53.90	30,800	10.92
1986	1,55,000	22.43	72,000	27.73
1988	2,22,000	32.12	70,800	26.47
1989	1,62,000	23.44	1,14,000	42.90
1990	34,648	5.01	15,307	5.22
1991	1,44,000	20.84	29,100	10.03
1993	2,92,745	42.36	37,469	12.84
1995	1,989.96	0.29	1,300.33	0.46
1996-97	10,500	1.52	3,592	1.31
2004	312844	45.27	67,534	22.34

Source: Revenue Department, Govt. of Assam.

Nf → No flood.

moderate though the area affected is quite extensive, in some other years crop damage is far more extensive though the area affected is somewhat less. This is because of the fact that the extent of crop damage caused by floods depends not only on the intensity but also on the timing of floods. Floods in the later part of a year is more devastating as it can cause damage to the standing kharif crops which is much more significant in importance in Barak Valley than the pre-kharif and the rabi crops. There was no extensive flood during the

Table 5
Damages Caused by Floods During 2004

District	*No of villages*	*Area af fected (in hect.)*	*Value of crop lost (in lakh)*	*House damaged (in lakhs)*	*Damages to public utilities (in lakhs)*
(1)	(2)	(3)	(4)	(5)	(6)
Cachar	434	178800	58.2	87997	3.2
Karimganj	175	121963	339.25	21315	778.5
Hailakandi	703	12081	–	2095	6904.78
Barak Valley	1312 (56.6)	312844 (45.27)	397.45	111407	7686.48
Assam	15950 (63.5)	2363906 (30.1)	37470.62	479337	17685.79

Source: Water Resource Dept., Govt. of Assam, 2004.

years 2006 and 2007, but untimely flood in the month of September, 2007, caused widespread damage to the crops and property of the people of the valley. The human suffering was also immense in that wave of flood. Though the exact estimates of value of the damage are not available at the moment, it is anticipated that the extent of damage surpassed the previous records. The Table-4 shows a comparative analysis of area affected and the extent of damage in the Barak Valley region and that of Assam. It is to be noted that the Brahmaputra and Barak together constitute 33.33% of water potential in India. Table-5 shows the damage caused by flood in Barak Valley in 2004 (Water Resource Department, 2004).

Irrigation Infrastructure

Irrigation is perhaps the most crucial component of agricultural infrastructure. The HYV fertiliser technology based strategy for agricultural growth through increase in crop yield and cropping intensity depends critically on irrigation. Though the Barak Valley region receives substantial amount of rainfall, the rainfall is not evenly

Table 6
Irrigation Potential Available and Irrigation Potential Utilised in the Barak Valley Region during the period 1990-91 to 2000-01

Year	*Irrigation potential available during the year (in hectares)*	*Irrigation potential utilised during the year (in hectares)*	*Estimated utilisation ratio (as percentage)*
(1)	*(2)*	*(3)*	*(4)*
1990-91	5610(1.91)	5066(1.73)	90.30
1991-92	7274(2.51)	4421(1.52)	60.78
1992-93	6726(2.28)	3533(1.20)	52.53
1993-94	7064(2.42)	698(0.24)	9.88
1994-95	5831(2.06)	1248(0.44)	21.40
1995-96	8024(2.83)	800(0.28)	9.97
1996-97	6299(2.23)	283(0.10)	4.49
2000-01	12380(4.09)	243.07(0.08)	1.96

Sources : Office of the Chief Engineer, Irrigation, Govt. of Assam.
Figures in parentheses are percentages of estimated total cropped area of respective years

distributed through out the year. About 70% of annual rainfall is usually concentrated in the summer months of May to September. This creates an imbalance in the natural water supply for cultivation. While excessive rain resulting in floods and waterlogging often cause damage to the kharif crop, the lack of rainfall in the winter months does not permit growing of rabi and pre-kharif crop extensively. To increase in cropping intensity and for reduction of instability in crop production through greater dependence on flood free crops, the development of irrigation infrastructure in the region in an imperative (Chatterjee, 1985).

Table-6 shows irrigation potential available and irrigation potential utilised in the Barak Valley during the period 1990-91 to 2000-01. A close look into the figures in the table shows a grim state of irrigation infrastructure in the region. Though irrigation potential available improved from 1.91% of estimated gross cropped area in 1990-91 to 4.09% in 2000-01, the ratio still remains extremely low to have any significant impact on agriculture. Moreover, the utilisation rate of this meagre irrigation potential created has fallen steadily over the years and utilisation rate was as low as 1.96% in 2000-01. Consequently, the area actually irrigated as a percentage of estimated total cropped area fell from 1.73% in 1990-91 to only 0.08% in 2000-01. It is to be noted here that the figures relating to irrigation infrastructure as mentioned in the table-4 relates only to government owned sources, mostly Lift Irrigation Schemes (LIS). For Barak Valley, the State Irrigation Department did not have the record for the privately owned irrigation systems. Indeed, such privately owned irrigation systems are practically non-existent in this region. If may be mentioned that in the Brahmaputra Valley of Assam there was some impressive expansion of privately owned irrigation systems during the decade of 1980s. As per statistics available from the Irrigation Department, Govt. of Assam, out the total irrigation potential available in the Brahmaputra Valley nearly a half was from privately owned systems. The growth of privately owned irrigation system in the Brahmaputra Valley was largely institutionally financed and the systems were installed by the public sector Assam State Minor Irrigation Development Corporation. These privately owned irrigation systems were mostly based on Shallow Tube Wells (STWs) which are suitable for exploiting the abundant ground water reserve at small depth. The absence of similar growth of ground water based privately owned irrigation systems in the Barak Valley might have been because of the non-availability of suitable ground water resource for similar exploitation in this region. As Das (1984) sums up, "The sediments comprising the water bearing horizons of the Barak Valley down to 50 metres are predominantly clayey and do not support good tubewells. It is therefore considered that tubewells in this region can not be successful." The recent

Table 7

Yield Rate of Paddy in Irrigated Farm(Kg per hectare)

	Early Ahu	Regular Ahu	Sali	Boro	Total Rice
(1)	(2)	(3)	(4)	(5)	(6)
Below 1 hectare	0	4357.46	5975.94	0	5436.45
1.0 to 2.0	0	5079.55	5434.57	4481.96	5325.69
2.0 to 4.0	0	4668.70	5996.95	0	5933.70
4.0 to 10	0	4481.96	5639.79	0	5617.09
Above 10 hectares	0	4780.75	5079.55	0	5060.88
Average	0	4673.68	5625.36	4481.96	5474.76

Source: Field Survey, 2009.

Table 8

Yield Gaps of Paddy in Irrigated and Unirrigated Farms

(Kg/Hectare)

	Regular Ahu	Sali	Boro	Total Rice
(1)	(2)	(3)	(4)	(5)
Irrigated Farms	4673.68	5625.4	4482.0	5474.8
Unirrigated Farms	1464.5	1584.5	1784.5	1611.0
Expected Gain/Loss(+/-)	(+) 3209.18	(+)4040.86	(+)2697.5	(+)3863.76

Source: Field Survey, 2009.

surveys by the Central Ground Water Board, Government of India, also confirmed this observation.

In our sample survey attempt has been made to collect information from the irrigated farms in the three districts of the Barak Valley. There are indeed a very limited number of irrigated farms in the valley. Accordingly, 30 samples have been collected from the irrigated farms in all the three districts. It is shown that the yield rates of the irrigated farms are much higher than the un-irrigated. It is to be noted that lack of irrigation along with incidence of natural calamities are the major constraints to agricultural development of the entire valley. From tables 7 and 8 it is seen that the expected gain due to irrigation facilities will be more than double the present productivity. The yield gap between irrigated and unirrigated farms indicates that with assured water supply in the paddy fields productivity can be enhanced in the Barak Valley. Thus the proposed project will be beneficial for the agricultural sector in the region.

Conclusions

Steps for development of irrigation infrastructure in the Barak Valley are urgently needed. However, for development of irrigation infrastructure, the region will have to depend on its river system for water, as it is not adequately endowed with quality ground water. Thus, emphasis should be made to develop some medium and large irrigation projects. The irrigation projects should be managed entirely by the farmers. A major project like the much awaited Tipaimukh River Valley Project appeared to be necessary for the purpose. However, since the project seems to have been abandoned by the Government of India due to the pressure from Bangladesh, an alternative flood control measure should be undertaken at the earliest. Raising and strengthening of the existing embankment to the required level is also urgently required. Increasing the navigability of the rivers is another important step which deserves attention. Water harvesting by using the wetlands within the valley should be given top priority for water resource management. Like Field Management Committees in the villages, the Water Users' Associations can also be popularized for sustainable use of water.

REFERENCES

Bhattacharjee, J. B. (1977), *Cachar Under British Rule in North East India*, Radiant Publishers, New Delhi.

Bhattacharjee, J.B. ed. (2009), *Development Strategies for Barak Valley (Assam)*, Akansha Publishing House, New Delhi.

Bezbaruah, M.P. (1997), 'Regional and Seasonal Variations in the

Trends in Rice Production in Assam from 1974-75 to 1994-95' (Summary), *Indian Journal of Agricultural Economics*, Vol.52, No.3, July-September, pp. 438.

Chatterjee, N., (1995), *Irrigated Agriculture*, Rawat Publications, Jaipur/Delhi.

Roy, N. & Bezbaruah, M.P. (2002), *Agricultural Growth and Regional Economic Development*, Mittal Publications, New Delhi.

Roy, Niranjan (2009), 'Socio Economic Survey of Barak Valley (THEP), A Report of the Major Project', Funded by NEEPCO, Assam.

Water Resource Department (2004), *A Report on Flood Damage in Assam*, Guwahati.

6

Sustainable Water Management in the Oxbow Lakes of Assam with Special Reference to Cachar District, Assam, Northeast India

Kankana Das & Susmita Gupta

[**Abstract:** The major sources of water in the rural areas of Barak Valley comprise ponds, oxbow lakes, temporary pools, beels, floodplain lakes, and other small aquatic systems. The three valley districts of Cachar, Karimganj and Hailakandi have 34 oxbow lakes occupying a total area of 717.50 ha. Oxbow lakes (locally called *anua*) are remnants of meandering floodplain rivers, which are physically isolated from the respective rivers and generally capture small relic drainages. Satkarakandi Anua, Phulbari Anua, Baskandi Anua, Bekirpar Anua and Ramnagar Anua are some of the major productive oxbow lakes in Cachar district. These are used as highly productive fisheries. Nearly 1200 ha area of these oxbows are used for major and medium irrigation projects as well. Water of these systems is also used for drinking and other domestic purposes by the inhabitants of nearby areas who are economically dependent on the *anuas*. Productivity and importance of and threats faced by the *anuas* along with some ameliorative measures are discussed in the paper.]

Introduction

Freshwater ecosystems play an important role in our lives. Important services provided by freshwater ecosystems include food, water for drinking, cooking and sanitation, materials for construction, production of subsistence and economic goods, flood control, and cultural services including spiritual, aesthetic, educational and scientific benefits. The majority of the world's population lives near the freshwater environment and depends on it variously, with most of the inland cities lying adjacent to a river or other large fresh water systems. In the developing world particularly, proper functioning of freshwater ecosystems have a proportionately greater impact on the livelihood, health and security of the poor.

India has about 4.1 million hectares of wetlands excluding paddy fields of which 1.5 million hectares are natural and 2.6 million are manmade. Wetland habitats in India have been consciously destroyed by draining and reclamation. Wetlands lost refers to physical loss and loss of 1 sq. km of wetland in India will have much greater impact than loss of 1 sq. km of wetland in low populated areas with abundant wetland (Foote *et al.* : 1996). In India, many wetlands of Bihar, Assam and West Bengal are of the oxbow type. Oxbows represent young types of reservoirs and are characterized by a very fast evolution. They are usually restricted to a littoral zone due to favorable light condition (Mikulski : 1974). Oxbow lakes are sections of meandering rivers that died down or changed their courses over time. These are connected with the floodplain of the river by inlets and outlets. By screening the inlets and outlets, an oxbow lake can be converted into a culture fishery. The name oxbow refers to its shape, which is like a horseshoe or oxbow. Oxbow lake falls into a category of common pool resources as it has the two characteristics of substractability and non-excludability of a common-pool resource (IFAD, DANIDA and UNOPS : 1994). The present paper throws light on the importance of the oxbow lakes of Cachar district, Assam, Northeast India, and the need for their sustainable management for the sustenance of human life in the region.

Oxbow lakes of Assam

In Assam a total of 861 ox-bow lakes/cut-off meanders are observed throughout the state covering an area of 15460.60 ha which constitutes 0.20 percent of the total geographical area of the state and 15.27 percent of the total area under wetlands (Garg *et al.* : 1998). The two major rivers of Assam are Brahmaputra and Barak, having 35 and 9 major tributaries, respectively. The river Brahmaputra, although flowing in a highly braided channel, forms large number of oxbow lakes throughout its course. The river enters the state in the Lakhimpur district in the east and flows out of the state through the Dhubri district in the west. The other major river Barak originating from the northern boundary of Manipur, and joined by river Jiri, meanders through the Cachar district of Assam forming a large number of oxbow lakes. Among all the districts of Assam highest number of oxbow lakes are recorded in Golaghat district (104), followed by Nagaon (71) and Dhubri (68) but area wise highest area under this category is observed in Morigaon district (2143 hactor) followed by Nagaon and Golaghat.

Oxbow lakes of Barak Valley

Barak Valley, situated in the southern part of Assam, is drained by river Barak which formed 34 oxbow lakes having a total area of 717.50 ha. The valley is comprised of three districts: Cachar, Karimganj and Hailakandi. The major sources of water in the rural areas of Barak Valley include ponds, oxbow lakes (*anua)*, temporary pools. flood plain lakes (*beels, haor*) and other small aquatic systems. Among the three valley districts, Cachar has the highest number (29) of oxbow lakes, followed by Hailakandi (4) and Karimganj (1), having an area of 592.50, 37.50 and 87.50 ha, respectively (Garg *et al.* : 1998). Oxbow lakes of Cachar district comprise 82.58% of the total oxbow lake area of Barak Valley. Total wetland area of Cachar district is 7188 ha and oxbow lakes constitute 8.24% of the total wetland area of the district. Satkarakandi Anua, Phulbari Anua, Baskandi Anua, Bekirpar Anua and Ramnagar Anua of Cachar district are some of the major productive systems of the district.

Table 1
Location, area and depth of a few oxbow lakes of Cachar district (as per 2009 report)

Sites	*Location (Distance from Silchar*	*Area*	*Depth*	*Area used by Irrigation Department, Silchar)*
1. Baskandi Anua	20 km	2.2 sq.km	4.2 m	180 ha (2008-09)
2. Satkarakandi Anua	13 km	1.4 sq. km.	4.2 m	120 ha (2008-2009)
3. Fulbari Anua	35 km	1.4 sq.km.	3.6 m	160 ha (2008-09)
4. Ramnagar Anua	2 km	1.2 sq.km.	4.2 m	160 ha (1992).
5. Bekirpar Anua	32 km	2.6 sq.km.	5.4 m	260 ha (2009-10)

Source: Department of Irrigation, Silchar Division, Assam, 2009.

Oxbow lakes and productivity

There are several studies which revealed oxbow lakes as highly productive habitats that are colonized by a variety of fishes during floods. For several species, oxbows appear to provide conditions that are more favourable for growth, survival, and reproduction than those associated with the river channel. Shallow oxbows are dominated by small, opportunistic species with attributes that are well-suited for rapid colonization of recently disturbed habitats (Winemiller and Rose : 1992). Oxbow lakes are nursery habitats and have positive influence on regional aquatic biodiversity. During high floods, oxbows and other floodplain depressions retain large volumes of water, thus decreasing the velocity and height of the flood over adjacent terrestrial landscapes (Sparks : 1993). Jhingran and Pathak (1987) estimated the fish production potential of the *beels* of Assam to be 1500 kg/ ha /year. According to Barua *et al.* (2000), the productivity of oxbow lakes can be increased to 2000-4000 kg/ha/year by introducing extensive or semi-intensive aquaculture.

In Barak Valley most of the fresh water systems play a vital role in the lives of many people providing sources of water, food, and employment. Oxbow lakes of Barak Valley are productive lakes rich in plankton, macroinvertebrates, macrophytes and variety of fishes. These fishes meet the protein demand of the inhabitants near the oxbows as well as in the outlying areas. Because of their crescent shape, turbulent nature of water, and connectivity with river, the oxbows harbour a variety of riverine fishes. In these lakes plankton constitute the major food for planktivorous fishes while aquatic insects form the essential food base for nesting and juvenile water fowls, predatory fish, and amphibians. Macrobenthic organisms occurring at the bottom of the reservoirs or lakes are also considered to be one of the important groups of fish food organisms, especially of bottom dwelling fishes (Manoharan *et al.*: 2006). Major macrophytes recorded in the oxbow lakes of Barak Valley are *Eichhornia crassipes* (Water hyacinth), *Chara* (musk grass), *Salvinia cucullata* (Flame moss), *Azolla pinnata* (Feathered mosquito fern), *Trapa bispinosa* (Water chestnut), *Jussiaea repens* (Primrose willow), *Cynodon dactylon* (Bermuda grass), *Nymphaea nouchali* (Blue water lily), *Nymphoides cristatum*, *N. indicum*, *Nelumbo nucifera* (water lotus) and *Hygrorhiza aristata* (Indian lotus). Important fish species found in the anua include: *Notopterus notopterus* (Kangla/Foli), *Channa punctatus* (Cheng/Goroi), *Channa marulius* (Gajar) *Channa orientalis* (Upal/Goroi), *Gudusia chapra* (Chapila), *Amblyphayngodon mola* (Mourola/Moka), *Heteropnystis fossilis* (Shinghi) , *Puntius ticto* (Punti), *Puntius puntius* (Punti), *Mystus vittatus* (Tengra), *Mystus tengara* (Tengra) etc. The *anuas* are famous for *Gudusia chapra and Amblyphayngodon mola.* They are used as highly productive fisheries by fishermen. Diversity of plankton, aquatic insects, and fishes of the said *anuas* indicate their productive nature.

Water quality of a few oxbow lakes (*anuas*) of Cachar

In the developing countries surface water bodies are utilized simultaneously as sources of drinking water, irrigation water, as sewage recipients, and as habitat for fish. As a result, inevitable conflict arises between these usages with grave consequences for the health and general welfare of the population (Renn *et al.*: 1992). Physico-chemical properties such as DO, free CO_2, alkalinity, pH, phosphate and nitrate of water of Baskandi Anua, Ramnagar Anua and Phulbari Anua were found to be within the respective permissible limits of the water quality standard of ISI (1982) and WHO (2004) except the microbiological property of surface water. For aquaculture, physico-chemical properties of water of the said *anuas* were found to be within acceptable concentration ranges.

Use of *anua* water for irrigation

The lands in the catchments of the *anuas* are used for agricultural purposes, and, hence, the Department of Irrigation, Silchar, Assam, is actively engaged in the establishment of many projects in these *anuas*. *Anuas* are used for water storage purposes as well; water retention and removal can only be carried out in accordance with, or subordinated to the interests of water management and agricultural water usage. So, for these purposes, the *anuas* are used for the construction of various major and minor irrigation projects. In the north-east region of India irrigated area is around 560,000 ha of which 69.8% area is in Assam (Department of Irrigation : 2009). At present, about 1100 ha of *anuas* are under the construction of different irrigation projects in Barak Valley. Among the said *anuas* the highest area used for irrigation by Irrigation Department, Government of Assam was recorded in Bekirpar Anua while lowest was recorded in Satkarakandi Anua (Table1). In Bangladesh portion of the Barak-Surma valley also oxbow lakes (locally called *Baor*) serve as valuable sources of irrigation water during the dry season.

Fish culture in Oxbow lakes of Barak Valley

In Barak valley, over the last two decades, when construction costs of traditional pond farms were rising, the fishery cooperatives operating over the *anuas* recognized the possibility of using the different *anuas* for a much more intensive fish production. Oxbow lakes or *anuas* in flood plains and outside protected river banks have basically different features and possibilities for fishery utilization. The *anuas* in the flood plains are connected to the main river bed for a shorter or longer period of time during the year. The development of an intensive pond-like production technology cannot, therefore, be undertaken in these water areas. The majority of oxbow lakes outside protected banks are used for intensive, pond-like fish production. About 90 percent of this area produces marketable fish from fingerlings by polycultural methods. About 10 percent of the area produces fingerlings through stocking. Such areas experience illegal capture of fishes at fingerling stages by the fish poachers, which need to be checked at the earliest for sustainable fish culture.

Discussion

Virtually all aquatic habitats experience periodic disturbances at some scale of space and time (Sparks et al. : 1990), and disturbances of human origin are increasingly influential. The lakes and reservoirs, all over the country without exception, are in varying degrees of environmental degradation. There has been a quantum jump in population during the last century without corresponding expansion of civic facilities resulting in lakes and reservoirs becoming sinks for incoming contaminants and leading to their impairment. A major challenge for fisheries managers is to gain a functional understanding of population and community response to environmental variation and disturbance at varying spatiotemporal scales.

Table 2
Productivity status of fish and types of fishery of a few oxbow lakes of Cachar district (as per 2009 report)

Oxbow lake	*Fish productivity/ Year (2009)*	*Types of Fishery*
Baskandi *Anua*	1250 kg/ha	Culture fishery
Ramnagar *Anua*	795 kg/ha	Culture fishery
Fulbari *Anua*	1200 kg/ha	Culture based
Bekirpar *Anua*	1100 kg/ha	Culture cum capture fishery
Satkarakandi *Anua*	800 kg/ha	Culture cum capture fishery

Source: Report of the Directorate of Fishery, Assam, Guwahati, 2009

Among the three districts of Barak Valley, while fish production increased in Hailakandi and Karimganj, in Cachar district it decreased from 17939 tons in 2006-2007 to 16950 ton in 2007-2008 (Bhagawati: 2009). Percentage of fish depletion rate was found to be highest in Cachar district at 5.78% / yr (Deka *et al.* : 2005). Fish yield trend and types of fishery in the oxbow lakes of Cachar district revealed that among the five *anuas* included in the present investigation, the highest productivity was recorded in Baskandi Anua (1250 kg/ha) followed by that in Phulbari and Bekirpar Anua (Table 2). In Baskandi Anua fish production decreased during 1990 to 2000 from 5060 kg / yr to 5000 kg / yr with a fish depletion rate of 1.18 % / yr. This showed that rate of fish production in 2000 was four times that of the present rate of production. In Satkarakandi Anua production decreased from 9262 kg/yr to 9200 kg/yr and depletion rate was 0.67%/yr. In Bangladesh a study recorded that fish yields of oxbow lakes have increased steadily during 1991-1995 from 137 kg/ha to 565 kg/ha (Middendorp *et al.*: 1996). The causative factors of depletion of fish in the wetlands of Assam are primarily degradation of physical environment and management deficiency (Deka *et al.*: 2005). A study made on the ecology of Baskandi Anua showed that the high rate of depletion was mainly due to infestation and spread of water hyacinth (Devi : 2010). There are records of prolific growth of water hyacinth in many lakes in India such as Loktak, Bhopal lakes, Ropar, Sukna, Kanjli, Pong Dam, etc. (Reddy and Char : 2004). The anuas of Barak valley are dominated by Indian Major Carps and *Gudusia chapra* (Chapila). They are followed by other miscellaneous fishes. In contrast, in the oxbow lakes of Bihar, the Indian Major Carps contributed less to the fishery, ranging from 3 to 22%. The miscellaneous fishes were found to be the main stay (Sinha and Jha : 1997). A study made by Das and Chakraborty (2006) in two tropical lakes has revealed that herbivores are very much efficient in energy utilization. In tropical lakes silver carp and grass carp were found to be most efficient in conversion of energy in the grazing and detritus food chains. Grass carp was found to be the best utilizer of food resources among all the species of fish cultured.

Regarding community based fisheries resource management, i.e. where the community themselves develop some norms and practices for the betterment of the fisheries resources as well as the security of their livelihood, in Assam the management strategy does not allow for the local fishing community to have a role in the management paradigm. Existing fishery cooperatives do not have meaningful function. The members are often ignorant about their rights and roles (Barua *et al.* : 2000). In Bangladesh, fisheries management of the oxbow lakes (*baors*) as community resource is done by the beneficiaries themselves. The lake management group is composed of a lake fishing team and a fish farming group. In the *baors* main parts of the lakes are managed as culture based fishery which include fingerling stocking, fish harvesting and regular de-weeding. *Baor* inlets and outlets are screened with bamboo, concrete or metal fencing (Middendorp *et al:* 1996).

Observations and Conclusions

The *anuas* of Cachar district are not maintained properly. The inhabitants being unaware of sustainable use of the aquatic systems use a large portion of surrounding areas as waste lands by throwing solid waste into the systems. They lack proper sanitation facilities rendering the microbiological quality of water of the *anuas* poor. The land nearby is very productive for agricultural practices throughout the year. However, use of pesticides and fertilizers in those lands has resulted in the accumulation of residues in the system through surface runoff, which has hazardous impact on the water quality and fauna and flora of the lakes. Further, encroachment for paddy cultivation has reduced the productive water surface of the lake. The combined effect of encroachment and cultivation has increased siltation in these wetlands reducing their fish production. Over-fishing, use of fine meshed fishing net, non-implementation of fishery rules are important causes of depletion of fishes. The extensive and uncontrolled growth of water hyacinth, their decomposition and silt load carried by rain water has reduced the depth of water which in turn affected migration of fishes from the river to the lakes. Another detrimental activity in Barak Valley that deserve mention is the rampant use of pesticides during dry season for trapping fishes, posing a hazard to human as well as ecosystem health.

Since the major area of the *anua*'s outside protected banks is used for intensive, pond-like fish production some technicai interventions and improvements are required. Existing fishery rules should be implemented and proper sanitation facility should be provided to the inhabitants by the competent authority. There is dire need of some environmental awareness or educational programme for the people living near these ecosystems.

[Acknowledgement

Authors are thankful to the Head, Department of Ecology and Environmental Science, Assam University, Silchar, for providing the laboratory facilities. We are also thankful to Mr. Dipak Das of Irrigation Department, Silchar and Mr. Sidhartha Purkayastha, Directorate of Fishery, Govt. of Assam, Guwahati, for their kind help.]

REFERENCES

Anonymous (2009), *Report on Irrigation scheme*, The Department of Irrigation, Assam, Silchar Divisional office, Nagatila, Silchar.

Anonymous (2009), *Report on Fishery development*, The Directorate of Fishery. Assam: Guwahati.

Baruah, U.K, A.K. Bhagowati, R.K. Talukdar and P.K. Saharia (2000), 'Beel Fisheries of Assam: Community-based Co-management Imperative', *NAGA*, The ICLARM Quarterly, 23 (2).

Bhagawati, P. (2009), *Fishery in Assam*, NEDFI Data bank.

Chakraborty, D. and S.K. Das (2006), 'Utilization and flow of energy in tropical ecosystem: A test case for two ox-bow lakes', *International Journal of Enviornmental Science and Technology*, 3 (2): 141-146.

Deka, T.K., M.M. Goswami and M. Kakati (2005), 'Causes of Fish Depletion- A Factor Analysis Approach', *NAGA: World Fish Center Newsletter* 28 (1 &2).

Devi, S.S. (2010), 'Ecology of Baskandi Anua, An Oxbow Lake Of Cachar District, Assam, North East India', Unpublished M. Phil dissertation, Assam University, Silchar.

Foote L., S. Pandy and N.T. Krogman (1996), 'Process of wetland loss in India', *Environmental conservation*, 23 : 45-54.

Garg, J.K., T.S. Singh, and T.V.R. Murthy (1998), *Wetlands of India*, Space Applications Centre (ISRO), Ahmedabad, India.

IFAD, DANIDA and UNOPS (1994), *Report on Oxbow Lake, Bangladesh*, Small Scale Fishermen Project: Mid Term Review.

ISI (1982), *Draft Indian standard specification for drinking water*, Indian Standards Institution, DOC: CDC 26, 8626, New Delhi.

Irrigation Department, Assam (2009), *The contribution made for creation of irrigationpotential, Irrigation Department and Assam* State Minor Irrigation Development Corporation, Guwahati.

Manohoran, S., V.K. Murugasan and R. Palaniswamy (2006), 'Numerical Abundance of Benthic Macroinvertebrates in selected reservations of Tamil Nadu', *J. Inland Fish. Soc. India* 38:54-59.

Middendorp, H.A J., R. Hasan and N.A. Ahmed (1996), 'Community fisheries management of Freshwater Lakes in Bangladesh', *NAGA*, The ICLARM Quarterly 19 (1).

Mikulski, J.(1974), *Biology of Inland Waters*, PWN, Warszawa, pp 434.

Reddy, M.S. and N.V.V. Char (2004), 'Management of Lakes in India: Theme paper in Lake management Initiative', India.

Renn, O., W.J. Burns., J.X. Kasperson, R.E.Kasperson and P. Slovic (1992), 'The social amplification of risk: theoretical foundations and empirical applications', *Journal of Social Issues* 48(4): 137-160.

Sparks, R. E., P.B. Bayley, S.L.Kohler and L.L. Osborne (1990), 'Disturbance and recovery of large floodplain rivers', *Environmental Management* 14: 699-709.

Sinha, M. and C. Jha (1997), *Ecology and Fisheries of Ox-Bow Lakes (Maun) of North Bihar - A threatened ecosystem*: Bulletin 74,

Central Inland Capture Fisheries Research Institute (ICAR), Barrackpore, West Bengal.

Winemiller, K.O. and K.A. Rose (1962), 'Patterns of life history diversification in North American fishes: implications for population regulations', *Canadian Journal of Fisheries and Aquatic Sciences* 49 : 2196-2218.

W.H.O. (2004), *Guideline for Drinking Water Quality*, 3rd Edition, World Health Organization, Geneva.

7

Hydrobiology of Salganga River, Cachar, Assam

Hafsa Sultana Laskar & Susmita Gupta

[**Abstract:** The study was undertaken in September 2006-August 2007 to analyze the physico-chemical properties of water and plankton ecology of Salganga river, an inlet of Chatla floodplain, Assam. The ecology of Salganga river is governed by flow seasonality imposed by monsoon rains. People residing near the river use its water for drinking, bathing and other domestic uses. The water level of the river goes down to several feet in dry months but at the onset of rainfall it becomes a part of Chatla floodplain. Low dissolved oxygen, high biological oxygen demand, slightly acidic pH and high nutrient (nitrate, phosphate, ammonia and sulphate) content of river water revealed that the water is not potable without treatment. The phytoplankton community was dominated by Chlorophyceae (green algae) while the zooplankton community was dominated by Cladocera. Shannon-Wiener diversity index-H' values were high for both phytoplankton (2.04) and zooplankton (1.8) and evenness (J') values were found to be almost similar for both the communities, i.e. 0.88 and 0.86 respectively. Rainfall, transparency and phosphate showed highly significant positive correlation with phyto and zooplankton density.]

Introduction

Water is essential for the development and maintenance of the dynamics of every facet of the society and thus availability of safe and reliable source of water is an essential pre-requisite for sustained development (UNSCD : 2000; Offem *et al.* : 2009). Running waters (rivers and streams) are perhaps the most impacted ecosystems on the planet as they have been the focus for human settlements and are heavily exploited for water supplies, irrigation, electricity generation, and waste disposal. As these systems have an intimate contact with their catchments, land-use alterations affect them directly (Malmqvist and Rundle : 2002). Hence limnological studies are necessary for evaluation of these stresses and to develop tools to convert scientific information into effective management decisions (Goldman : 2003).Various groups of plankton (algae, diatoms and zooplankton) are often used in the routine water quality monitoring in various aquatic ecosystems, as biomonitoring is a central component of water resource management worldwide (Rosenberg and Resh : 1993).

Salganga river originating from the foothills of Mizoram flows by the village Salganga, Silcoorie and finally flows through the Chatla Floodplain (*haor*) of Cachar district, Assam (Fig.1). During the prolonged monsoon inundation, the river becomes a part of the floodplain. During the process, it receives variety of migratory fishes that comes to Chatla floodplain for spawning and breeding. The riparian vegetation of the river include *Barringtonia acutangula* (Hijol), *Albizzia sp.* (Shirish), *Bombax spp.* (Simul), *Lagerstroemia flosreginae* (Jarul), *Zizipus sp.* (Kul), *Mangifera indica* (Mango), and Bamboo clumps (various species of *Bambusa*), cluster of *Calamus tenuis* (Rattan), etc. The river water has submerged, emergent and floating macrophytic vegetation like *Vetiveria zizanoids* (Tattoo grass), *Juncus sp.* (Hedgehog Grass/ Rush-weed), *Cyperus sp.* (Sedge- local name: Manu grass), *Polygonum sp.* (Knotweed - local name: Khukra), *Trapa sp.* (Shingai), etc. People use river water for drinking, bathing and other domestic purposes. In the deep-water zone part time fishing is also a regular practice. As the river caters to the needs of the people of Salganga and Silcoorie, and it is also one of the inlets to a major floodplain of Barak Valley, the present study aims to evaluate the human impact on the ecology of the river using plankton community as bioindicator. The study also aims to investigate the impact of inundation during monsoon on the

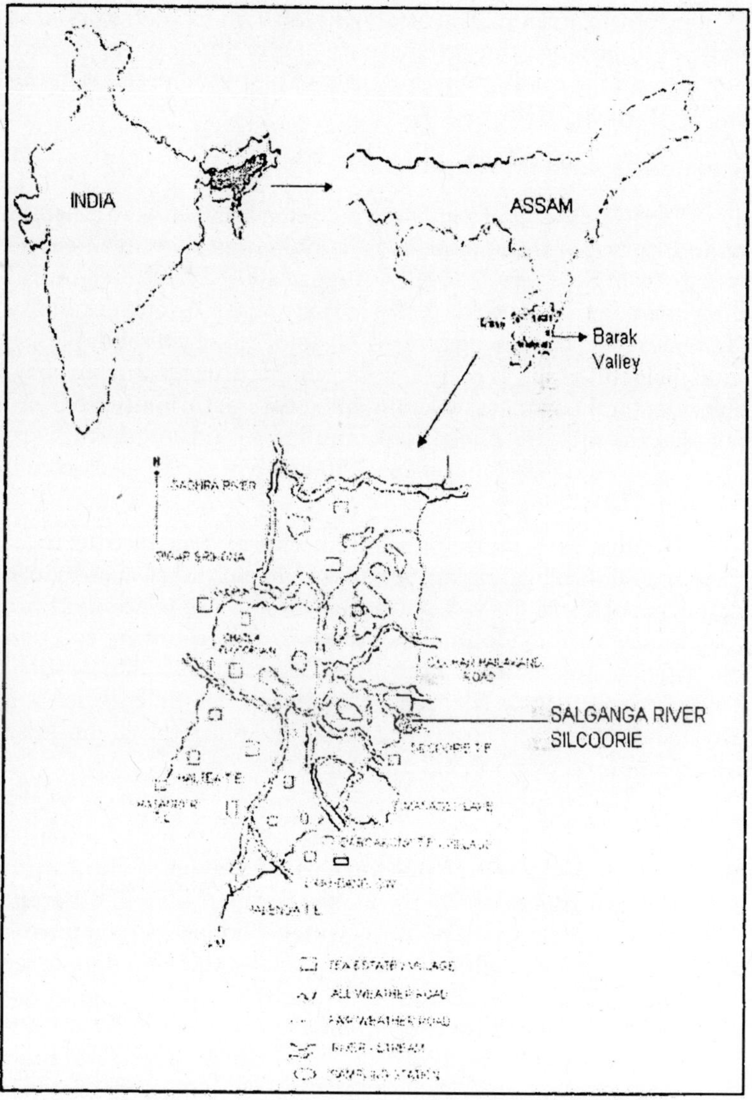

Fig. 1: Map of Salganga river, Cachar, Assam showing the location of sampling station

ecology of the stream The study of plankton in such system is important because the productivity of any aquatic water body depends on the quality and quantity of plankton present in the said water body (Guy : 1992).

Materials and Methods

Three replicates of water and plankton samples were collected at an interval of three months from Salganga river at Silcoorie village from September 2006 to August 2007. Physico-chemical properties of water such as dissolved oxygen (DO), total alkalinity (TA), pH, electrical conductivity (EC), total dissolved solids (TDS), total suspended solids (TSS), water temperature, transparency, chloride, total hardness, calcium, magnesi·:m, turbidity, sodium, potassium, nitrate, phosphate, ammonia and sulphate were analyzed by standard methods (Golterman *et al.*: 1978; Michael: 1984; APHA: 2005).

Quantitative sampling of plankton was done by filtering a known volume of water through plankton net (mesh size 40μm). After fixation in 4.5% per cent formalin and pre-concentration, cell counts were made in a 'Sedgwick Rafter' counting cell and identified using standard literature (Edmondson: 1959; Battish: 1992; Anand: 1998). Pearson correlation coefficients among physico-chemical parameters of water and plankton community were computed using SPSS 12.0 for Windows.

Results

(i) **Physico-chemical properties of water of Salganga river:** Table 1 depicts the mean concentrations of different physico-chemical variables of water of Salganga river during 2006-2007. Rainfall ranged from 31.1-470.1 mm and water temperature ranged from 18.67-32.67°C. Both transparency and turbidity showed wide ranges of variations from 3.42-10.8 cm. and 1.83-12.56 NTU, respectively. The DO of river water was found to be low with a narrow range of variations (2.64-5.26 mg l^{-1}) whereas BOD was consistently high ranging from 6.6-12.06 mg l^{-1}. TDS and pH ranged from 22.52-28.6 mg l^{-1} and 5.6-6.74, respectively. Concentrations of chloride, nitrate and phosphate ranged from 7.7-91.4 mg l^{-1}, 2.95-4.87

Table 1

Physico-chemical properties of water of Salganga River during September 2006 to August 2007

Parameters	*Range*	*Mean ±SE*
DO (mgl $^{-1}$)	2.64-5.26	3.68±0.57
T.Alkalinity (mgl $^{-1}$)	46.6-74.91	62.84±6.83
pH	5.6-6.74	6.27±0.25
Conductivity (µs cm $^{-1}$)	3666.67-4574	4164.34±224.61
TDS (mgl $^{-1}$)	22.52-28.6	24.98±1.34
TSS (mgl $^{-1}$)	15.8-50.9	27.49±8.23
Free CO_2 (mgl $^{-1}$)	6.98-14.08	9.6±1.62
Water Temperature (^{0}C)	18.67-32.67	27.51±3.08
Transparency (cm)	3.42-10.8	6.54±1.55
Chloride (mgl $^{-1}$)	7.7-91.4	33.85±19.56
T.Hardness (mgl $^{-1}$)	17.7-58.3	34.12±8.95
Calcium (mgl $^{-1}$)	13.2-40.9	23.96±6.47
Magnesium (mgl $^{-1}$)	4.52-17.4	10.17±2.67
BOD_5 (mgl $^{-1}$)	6.6-12.06	8.31±1.28
Sodium (mgl $^{-1}$)	3.5-6.0	5.01±0.54
Potassium (mgl $^{-1}$)	1.42-1.9	1.59±0.16
Turbidity (NTU)	1.83-12.56	6.57±2.23
NO_3 (mgl $^{-1}$)	2.95-4.87	3.81±0.4
PO_4 (mgl $^{-1}$)	0.83-9.66	4.41±2.04
NH_3 (mgl $^{-1}$)	0.23-0.83	0.57±0.15
$S0_4$ (mgl $^{-1}$)	0.93-2.56	1.95±0.35
Rainfall (mm)	31.1-470.1	218.88±96.96

mg l^{-1} and 0.83-9.66 mg l^{-1} respectively. **Ammonia** and sulfate concentrations ranged from 0.23-0.83 mg l^{-1} and 0.93-2.56 mg l^{-1}, respectively.

(ii) **Plankton community composition :** Phytoplankton community of the river was represented by five major groups, viz. Chlorophyceae (green algae), Cyanophyceae (blue-green algae), Bacillariophyceae (diatoms), Euglenophyceae (euglenoids) and Dinophyceae (dinoflagellates). Chlorophyceae was found to be the most dominant group in terms of density followed by Cyanophyceae, Bacillariophyceae and Euglenophyceae. Dinophytes were the least dominant group throughout the study period. Zooplankton community was represented by Cladocera, Rotifera, Copepoda, Brachiopoda and Rhizopoda. The most dominant group was Cladocera in terms of density followed by Copepoda and Rotifera and Brachiopoda. Record of Rhizopods in the system was negligible throughout the study period (Fig. 2).

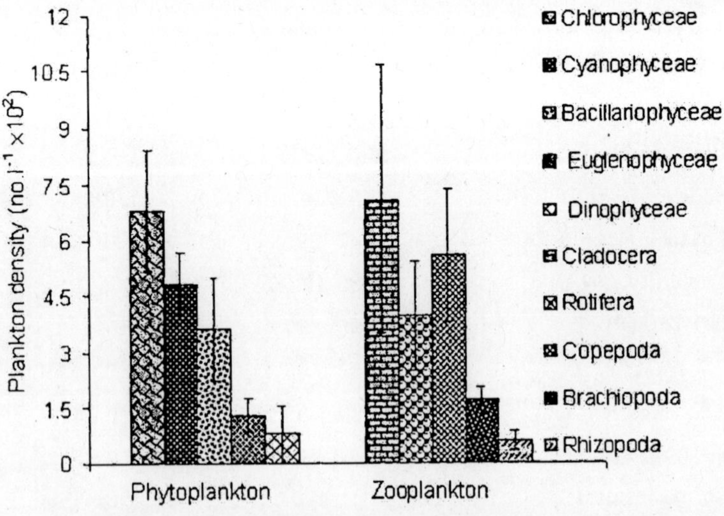

Fig. 2: Density of different groups of phytoplankton and zooplankton community of Salganga river during September 2006 to August 2007

(iii) Community Structure : Shannon-Wiener diversity index-H′ values were high for both phytoplankton (2.04) and zooplankton (1.8) and evenness (J′) values were found to be almost similar for both the communities, i.e., 0.88 and 0.86 respectively (Table 2).

Table 2
Diversity indices of phyto and zooplankton community of Salganga River during September 2006 to August 2007

Diversity Indices	*Shannon-Wiener index (H′)*	*Evenness Index(J′)*
Phytoplankton	2.04±0.2	0.88±0.04
Zooplankton	1.8±0.07	0.86±0.02

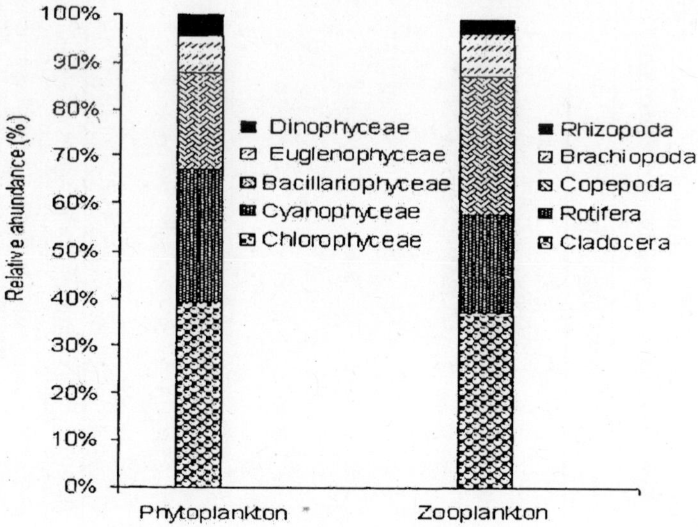

Fig. 3: Relative abundance of different groups of phytoplankton and zooplankton community of Salganga river during September 2006 to August 2007

Table 3

Pearson correlation coefficient matrix among different physico-chenical variables, phytoplankton density and zooplankton density of Salganga River

Parameters	*DO*	*TA*	*pH*	*EC*	*TDS*	*WT*	*TRNS*	Cl_2	*BOD*	NO_3	PO_4	NH_3	*PHYTO*	*ZOO*	*RF*
DO	-	0.72**	0.78**	0.47	0.33	-0.71**	-0.24	0.88**	-0.16	-0.4	0.13	-0.68*	0.02	-0.29	-0.63*
TA		-	0.51	0.65*	0.27	-0.58	-0.69*	0.64*	0.48	0.09	-0.48	-0.13	-0.49	-0.70*	-0.80**
pH			-	0.28	0.04	-0.43	-0.12	0.72**	0.14	-0.10	0.26	-0.48	0.14	-0.19	-0.49
EC				-	0.74**	-0.28	-0.44	0.36	0.29	0.12	-0.15	-0.27	-0.06	-0.16	-0.17
TDS					-	-0.46	0.01	0.39	-0.38	-0.42	0.28	-0.60*	0.19	0.147	0.16
WT						-	0.35	-0.91**	0.3	0.65*	0.04	0.58*	0.26	0.50	0.59*
TRNS							-	-0.28	-0.65*	-0.39	0.83**	-0.36	0.64*	0.77**	0.64*
Cl_2								-	-0.22	-0.55	0.11	-0.68*	-0.10	-0.44	-0.66*
BOD									-	0.83**	-0.69*	0.61*	-0.44	-0.42	-0.36
NO_3										-	-0.49	0.76**	-0.16	-0.14	0.02
PO_4											-	-0.71**	0.85**	0.76**	0.48
NH_3												-	-0.50*	-0.28	0.06
PHYTO													-	0.86**	0.57*
ZOO														-	0.82**
RF															-

** Correlation is significant at the 0.01 level * Correlation is significant at the 0.05 level.

DO= Dissolved oxygen, TA= Total alkalinity, EC= Electrical conductivity, TDS= Total dissolved solids, TSS= Total suspended solids, WT= Water temperature, TRNS= Transparency, Cl_2= Chloride, BOD= Biological oxygen demand, $N0_3$= Nitrate, PO_4= Phosphate, NH_3 = Ammonia, PHYTO= Phytoplankton density, ZOO= Zooplankton density, RF= Rainfall

(iv) Correlation Analysis : The Pearson Correlation Coefficient Analysis revealed that DO has highly significant positive correlation with alkalinity, pH, chloride and significant negative correlation with water temperature, ammonia, rainfall and BOD. Rainfall, transparency and phosphate showed highly significant positive correlation with phyto and zooplankton density whereas alkalinity showed highly significant negative correlation with zooplankton density. Ammonia showed highly significant positive correlation with BOD, nitrate and significant negative correlation with TDS, phosphate and phytoplankton density (Table 3).

Discussion

Rivers integrate all that happens in their landscapes, with their condition, especially their biological condition, revealing much about the consequences of human actions. The low DO, high BOD and slightly acidic pH of Salganga water reveal that it is not of potable quality without adequate treatment. A study made in the lotic systems of Tamil Nadu also reported low DO and high BOD in the downstream of the lotic systems (Drusilla *et al.*: 2004). Wide range of concentrations of chloride, phosphate, and sulphate depicted impact of rainfall. Inverse relationship of DO with temperature, BOD and rainfall, and significant positive relationships of ammonia with BOD and nitrate clearly indicate that surface runoff transported sewage, fertilizer etc. into the river which in turn increased the BOD. This increase, coupled with high temperature resulted in the lowering of DO. High BOD of the river water could also have been contributed by the entry of human and animal excreta as reflected in the poor microbiological water quality of ponds, streams, wells, and other drinking water sources of Barak Valley (Gupta : 2004; Gupta *et al.* : 2008; Saikia : 2009). Laskar and Gupta (2009) also found inverse relationships of DO with BOD, water temperature and rainfall in Chatla floodplain. Availability of nutrients such as nitrogen, phosphorus and sulphate concentration in water plays an important role in determining phytoplankton community productivity levels (Yamamuro *et al.*, 1993). River water was found to be dominated by Chlorophyceae (39.21%) in terms of density (Fig. 2 and 3). This is a common feature of lotic systems with flowing water (Uttah *et al.*: 2008). However, density of two other groups, viz., Cyanophyceae (27.88%) and Bacillariophyceae (20.91%) in the river water was also prominent. Bacillariophyceae are often dominant in rivers as these genera favour a turbulent environment with plenty of water mixing accompanied by low light conditions (Moss *et al.* : 1989). Presence of Euglenophyceae (7.45 %) and Dinophyceae (4.56 %), although in low proportions indicated a nutrient rich environment as has also been shown by Nwankwo and Amuda (1993) in the lower reaches of Okpoka Creek, Nigeria. Previous studies in the Chatla floodplain also reported that *Euglena* blooms were found to be induced by high concentrations of ammonia - nitrogen, nitrate - nitrogen, iron, magnesium and to some extent, phosphate copper and zinc in their water (Duttagupta *et al.*: 2004; Bhuiyan and Gupta : 2007). Zooplankton species are cosmopolitan in nature inhabiting all freshwater habitats of the world, including polluted industrial and municipal wastewaters. They are not only useful as bioindicators to help us detect pollution load, but are also helpful for ameliorating polluted waters (Mukhopadhay *et al.* : 2007). Zooplankton community of the river was found to comprise Cladocera (37.43%), Copepoda (29.51%), Rotifera (20.95%), Brachiopoda (9.04%) and Rhizopoda (3.08%) (Fig. 3). In the zooplankton population of river water Cladocera was found to be the most dominant group followed by Copepoda and Rotifera while Brachiopoda and Rhizopoda were found to be the least abundant. This agreed with the finding of previous study made in a marsh of the Chatla floodplain (Laskar and Gupta : 2010).

[**Acknowledgements**: Authors are thankful to the Head, Dept. of Ecology & Environmental Science, Assam University, Silchar, Assam for providing laboratory facilities. The first author is thankful to UGC, New Delhi, for financial support.]

REFERENCES

Ananda, N. (1998), *Indian Freshwater Microalgae*, Bishen Singh Mahendra Pal Singh, Dehradun, India.

APHA (2005), *Standard Methods for the Examination of Water and Wastewater*, 21st edn., APHA, AWWA, WPCP, Washington DC, USA.

Battish, S.K. (1992), *Freshwater zooplankton of India*, Oxford and IBH Publishing Co. Pvt. Ltd.

Bhuiyan, J.R. and S. Gupta (2007), 'A comparative hydrobiological study of a few ponds of Barak Valley, Assam, and their role as sustainable water resources', *J. Environ. Biol.*, Vol. 28, No. 4, 799-802.

Drusilla, R., A. Kumaresan and M. Narayanan (2004), 'Studies on water quality parameters of lotic systems in and around Courtallam (Tamil Nadu)', *Poll Res.*, Vol. 23, No. 3, 515-521.

Duttagupta, S., S. Gupta and A. Gupta (2004), 'Euglenoid blooms in the flood plain wetlands in Barak Valley, Assam, North Eastern India', *J. Environ. Biol.*, Vol. 25, No. 3, 369-373.

Edmondson, W.T. (1959), *Fresh water Biology*, 2nd edn., John Wiley and Sons, New York.

Guy, D. (1992), *The ecology of the fish pond ecosystem with special reference to Africa*, Pergamon Press, pp. 220-230.

Golterman, H.L., R.S. Clymo and M.A.M. Ohnstad (1978), *Methods of Physical and Chemical Analysis of Freshwaters*, 2nd edn., Blackwell Scientific Publishers, Oxford, London.

Goldman, C. (2003), 'Strategies for lake management in an increasing global environment, pp. 177-190', In Kumagai M. & Vincent, W.F. (ed.), *Freshwater management : global versus local perspectives*, Springer, Stuttgart, 233p.

Gupta, A. (2004), "Drinking Water Quality in Barak Valley, Assam, North East India : Prioritizing Management Options', In *Proceedings of the 2nd APHW Conference*, Vol. I, pp. 621-627, Asia Pacific Association of Hydrology and Water Resources, Japan.

Gupta, S., K. Buragohain and J.R. Bhuiyan (2008), 'Importance of ponds in integrated water resource management in Barak Valley, Assam, India', *Poll. Res.*, Vol. 27, No. 3, 587-590.

Laskar, H.S. and S. Gupta (2009), 'Phytoplankton diversity and dynamics of Chatla floodplain lake, Barak Valley, Assam, North East India – A seasonal study' *J. Environ. Biol.*, Vol. 30, No.6, 1007-1012.

Laskar, H.S. and S. Gupta (2010), 'Ecology of a marsh in Chatla floodplain, Barak valley, North-East India', *Eco. Env. & Consv.* Vol. 16, No. 3, 9-15.

Malmqvist, B. and S. Rundle (2002), 'Threats to the running water ecosystems of the world', *Environ. Conserv.*, Vol. 29, No. 2, 134–153.

Michael, P. (1984), *Ecological Methods for Field and Laboratory Investigations*, Tata Mc Graw-Hill publishing Company Ltd. New Delhi pp. 434.

Moss B., I. Booker, H. Balls and K. Manson (1989), 'Phytoplankton distribution in a temperate floodplain lake and river system. I. Hydrology, nutrient sources and phytoplankton biomass', *J. Plankton Res.*, Vol.11, 815-840.

Mukhopadhyay, S.K., B. Chattopadhyay, A.R. Goswami and A. Chatterjee (2007), 'Spatial variations in zooplankton diversity in waters contaminated with composite effluents', *J. Limnol.*, Vol. 66, No. 2, 97-106.

Nwankwo, D.I. and S.A. Amuda (1993), 'Periphyton Diatoms on three floating Aquatic Macrophytes in a polluted South-Western Nigerian Creek', *Intern. J.Ecol. Environ. Sci.*, Vol.19, 1-10.

Offem, B.O., Y.A. Samsons, I.T. Omoniyi, and G.U. Ikpi (2009), 'Dynamics of the limnological features and diversity of zooplankton populations of the Cross River System of SE Nigeria', *Knowl. Managt. Aquatic Ecosystem*, Vol. **393**, No. 02, 1-19.

Rosenberg, D.M. and V.H. Resh (1993), *Freshwater biomonitoring and benthic invertebrates*, New York: Chapman and Hall. 448p.

Saikia, K. (2009), 'Assessment of surface water quality and quantity in selected villages of Cachar and Karimganj district of Assam'. Unpublished M Phil dissertation, Assam University, Silchar

United Nations Commission on Sustainable Development (UNCSD) (2000), *Comprehensive assessment of the Freshwater resources of the World*, UN, New York.

Uttah, E.C., C. Uttah, P.A. Akpan, E.M. Ikpeme, J. Ogbeche, L. Usip and J. Asor (2008), 'Bio-survey of Plankton as indicators of water quality for recreational activities in Calabar River, Nigeria', *J. Appl. Sci. Environ. Managment.*, Vol.12, No. 2, 35 – 42.

Yamamuro, M., I. Kioke and H. Iizuumi (1993), 'Partitioning of nitrogen stocks in the vicinity of a Fijian seagrass bed dominated by *Syingodiun isoettifolum* (Ascherson) Dandy', *Australian Journal of Maine and Freshwater Resource*, Vol. 44, 101-115

8

Rainwater Harvesting for Sustainable Agriculture in Barak Valley

Mithra Dey & Nikhil Bhusan Dey

[**Abstract:** Water is the essence of life, where as water resources are fast depleting or becoming unsuitable for human use. The UN report published before the Third World Water Forum in Kyoto (Japan) in 2003 pointed out that world water reserves are fast drying up and that this fact accompanied by rising population, pollution and global climate change will bring about severe water crisis in near future. Agriculture and its allied activities like fishery forms the economic backbone of rural India and more so in the Northeast where fish is a highly preferred food item. The Barak Valley of South Assam has a unique topography and is endowed with abundant natural water resources. The River Barak and its tributaries along with the wetlands can provide abundant scope for sustainable agriculture and fishery development. Due to poor irrigation facilities agriculture suffers when rainfall is insufficient and irregular. Declining rainfall in the valley along with the rest of country is also a major concern. On the other hand, excess and sudden heavy showers lead to water logging and flash floods which can be managed by proper planning. Rainwater harvesting, proper management of existing water resource, watershed development and community participation will help attain sustainable agriculture and uplift the socioeconomic conditions in the rural sector. This paper studies various constraints of water management in agriculture sector and puts forward some suggestions to achieve sustainability in agriculture and allied activities in Barak Valley through water resource management. It also tries to emphasize the role NGOs, SHGs and cooperative societies can play in this direction.]

Introduction

History bears evidence to the fact that all civilizations developed on the banks of rivers and water is the basic necessity of life. One cannot imagine a day without water. More than 5 million people die every year from water related diseases. Fresh water is available from rivers, ponds, lakes; ground water including soil moisture, swamp water and permafrost and the rest is trapped in glaciers and permanent snow cover. Salt water or saline water is trapped in oceans and seas and forms 97.5% of global water estimates. Fresh water is only 2.5 % and of this 0.3 % is in lakes and rivers; 30.8 % is ground water and 68.9% is in glaciers and permanent snow cover.

Water is not an internationally tradable commodity which can be purchased or acquired from one part of the globe and sold in another. People need water sources close to their dwelling place for their daily requirement. The main supply of water is for domestic purposes, agriculture and irrigation and allied activities like fishery, industry etc. which comes from the surface water (rivers and lakes) and/or from ground water. 60% of the total land cultivated in India depends on groundwater for irrigation. Groundwater is also used for drinking. But the availability as well as quality of the groundwater cannot be taken for granted, because the overexploitation of groundwater generally leads to receding levels, degradation of water quality and finally, water scarcity. A recent report states that 33% of groundwater in India is unfit for drinking purposes as it is contaminated with fluoride, iron and arsenic and in some parts of the country salinity is another problem (TOI, March 12th, 2010). Similar contamination of the ground water by arsenic has been reported from the Ganga-Brahmaputra-Meghna basin which includes West Bengal and Assam as well as Bangladesh (Gupta et al, 2008). Such contamination has prompted the region to think about utilizing more of the available surface water and rainwater harvesting which can be achieved both by traditional methods as well as modern techniques. The U N report published before the IIIrd World Water Forum in Kyoto, Japan (16th-23rd March, 2003) pointed out that world water reserves are fast drying up and that the rising population, pollution and global warming will create terrible water scarcity in the next 20 years. Agriculture is the backbone of Indian economy and it utilizes the bulk share of water resources available in the country. Agriculture at present uses 524 billion cubic meters of water which is expected to rise to 807 billion cubic meters by 2050.

In the north-eastern states of India, agriculture and its allied activities forms the main occupation of the people, either in the plains or in the hilly terrains of the region. Agriculture and its allied activities occupy a vital position in the economy of Assam, engaging about 70 percent of the total workforce and the state has the inherent strength to attain food self-sufficiency. Food crops dominate the entire cropping pattern in Assam and mainly include rice, vegetables, oilseeds and pulses. The mainstay of economy in the Barak Valley of Assam is also agriculture. Like the rest of India, the cropping pattern in the Barak Valley districts is determined by natural factors like rainfall, climate and soil conditions. Irrigation holds the key for increasing the productivity and cropping intensity in any area. Although three departments of the Government of Assam are entrusted with the development of irrigation facilities the progress seems to be far from satisfactory.

Study Area

The present study highlights the water resource scenario in the Barak Valley of South Assam keeping in mind the agriculture potential of the valley. The Barak Valley comprises the districts of Cachar, Karimganj and Hailakandi and is situated between 24° 8'N – 25° 8'N and 92° 15' E – 93° 15'E. The valley, drained by the Barak River and its tributaries, is bounded by the North Cachar Hills District of Assam and the Jaintia Hills District of Meghalaya in the north, Mizoram in the south, Manipur to the east and the state of Tripura and Sylhet districts of Bangladesh on the west. The valley abounds in wetlands in the floodplains of the Barak River and its tributaries; tanks and ponds of various sizes, streams, pools, marshes and others wetlands. It is gifted with large myriad riverine tectonic lakes which bear the potential of supporting agriculture as well as fishery. It has large number of ponds which are used to meet water requirements and also to rear fishes of different varieties. These ecosystems help in storing the excess water after heavy rainfall in the monsoons. The River Barak flows a distance of 192 kms in the valley receiving a large number of tributaries, viz., Jiri, Chiri, Badri, Madhura, Jatinga, Hareng and Kalaincherra on the north and the Sonai, Rukni, Ghagra and Katakhal on the south. The river itself bifurcates into the Surma, flowing northwards, and the Kushiara, flowing southwards, at Haritikar near Badarpur. Longai and Singla are the important tributaries of Kushiara, while Ghumra is the most important tributary of Surma. All these rivers, especially the Barak which adopts a meandering path through the alluvial plains, frequently shift their courses and form ox-bow lakes. Wetlands occupy a total area of 13737.5 ha, of which 10016 ha is seasonally flooded.

The Barak Valley region has a subtropical climate. Warm and humid climate generally prevails during the summer (July/ August). During rainy season, the air is surcharged with moisture and rainfall is extremely heavy. The average annual rainfall of the region is c2500mm. The humidity ranges from 65-70 during winter to 85-90 percent during rainy season. The rainy season starts from May and continues up to October. Winter is short and mild (December to February). The months of March and April witness severe storms (locally called *kaal-vaishaakhi*) with strong wind, thunder and hail. Table 1 represents the pattern of rainfall in Barak Valley.

Table 1: Rainfall Pattern in Barak Valley

Between June-September	61%	Monsoon
Between March to May	30%	Pre-monsoon
Between October to February	9 %	Winter.

The monthly rainfall varies between 0.0 mm in December-January to around 439-687mm in the peak rainy season, i.e. May to August. However, the average annual rainfall from 1997-2000 and 2004-2008 shows a decreasing trend (Fig. 1). The pattern of rainfall in recent years has shown a change, with more spells of

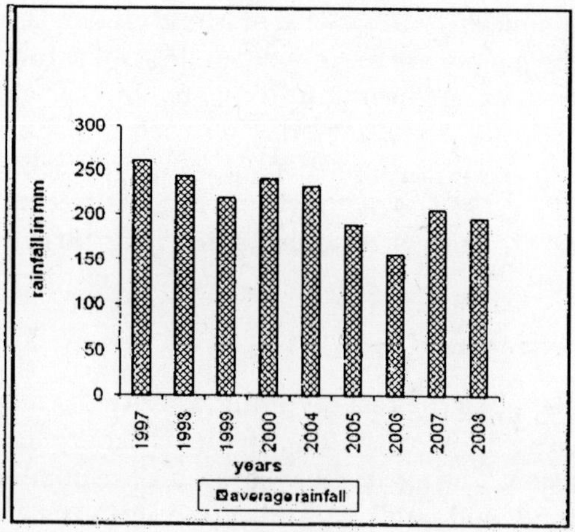

Fig.1. Average Annual Rainfall (mm)

dry days and fewer days of heavy rainfall leading to drying up of shallower habitats and shrinking of wetlands and other water bodies. Flash floods have also become more common. According to the South Assam Meteorological Subdivision (that covers the hills of Nagaland, Mizoram, Manipur and Tripura and parts of Borail Hills in South Assam), a significant change in seasonal rainfall has been observed. The summer monsoon rainfall is found to be decreasing over the region significantly during the last century, approximately 11mm per decade [Das, 2004; Mirza et al, 1998]. Like Barak Valley region, several districts of Assam have also experienced drought like conditions in recent years [IPCC, 2007a]. Similar change in the climate pattern in other parts of Northeast India has also been reported [Das, 2009].

The soil of the region varies from alluvial to laterite soil. Texture is generally clay loamy to clay. The pH ranges from 4.5 to 6. The riverine tracts are found to be loamy to sandy loam soil. These soils are less acidic in reaction and pH ranges from 5.00 to 6.00. The hilly tracts are generally predominated by laterite soils. These types of soil are acidic in nature due to heavy precipitation.

A survey conducted on a few villages in and around Dargakona revealed that the people are mainly dependent on rainwater for agriculture. They do have community wells, but for obtaining clean water for daily activities they do resort to harvesting rainwater. However, the system adopted is not sufficient to meet the requirements as they do not have any large collecting facility that would help in time of drought or for long term rainless period.

Agricultural Scenario

Barak valley, like the rest of Assam, is virtually dependent on agriculture and allied activities, of which fishery occupies a significant position. The agricultural systems are dominated by peasant subsistence and small scale mixed family farms. However, the agricultural productivity is far from satisfactory and the practices being more traditional the yield is very low. Cropping pattern in the area includes cultivation of paddy, followed by rapeseed and mustard, potato, arecanut, banana, sugarcane and chillies. Jute and tea plantation occupies important place in the economy of this region. Paddy is the most important crop, which includes *Ahu* paddy in summer, followed by *Sali* paddy. *Boro* paddy is grown in the low lying areas; *Ahu* paddy is followed by winter vegetables, oilseeds and pulses. Summer vegetables are grown in riverine tracts.

"Irrigation increases yields, permits new seeds to be used and adds greatly to the reliability of harvests." (Maddison, 1970). Irrigation holds the key for increased productivity and crop intensity. At present, three Govt. Departments, viz. Irrigation, Agriculture and DRDA are associated with development of irrigation facilities in the region but irrigation facilities in reality are far from satisfactory. The progress in extending irrigation facilities to the farmers are found to be very limited mainly due to: financial crunch, power shortage, internal disputes among the villagers, lack of initiative from the cultivators which prevent the implementation of various schemes. 80% of the existing schemes suffer from improper maintenance and poor coverage. Presently, Cachar district is having a cropping intensity of only 13.0 percent per hectare of land. This means that most of the cultivable lands remain fallow and is not cultivated. This colossal waste is indeed a routine picture of the vast landscape in the valley from December to April due to lack of irrigation facilities (Dept. of Agriculture, Govt. of Assam). Despite the availability of abundant surface and ground water resources, large section of cultivable land is found lying fallow. This land could be successfully utilized for multiple cropping and for developing fish culturere. Although financial crunch is said to be a cause of poor irrigation, according to a report published in a leading newspaper very recently the Government of India has sanctioned substantial amount to the Government of Assam for the development of irrigation infrastructure under the Auxiliarated Irrigation Beneficiary Project. However, the benefits of the project are yet to reach the agriculturists in the Barak Valley.

Fisheries Resource

Assam is the most water resourceful state in Northeast India with an estimated coverage of 3.74 lakh hectares of water spread areas, as per information available from the State Directorate of Fisheries. However, present level of annual fish production in the state is 1.81 lakhs MT against estimated demand of 2.83 lakhs MT, calculated on the basis of minimum nutritional requirement of 11 Kg per capita annually. The production rate is much lower than the potential capacity of the resources available. The gap is partially met by importing fish from other states. The present per capita consumption is about 7.3 Kg annually. In Barak Valley the annual demand for fish is approximately 62 MT, but the production is about 46-50 MT annually. The area is rich in fishery resources in the form of *beels*, lakes, rivers, ponds, forest fisheries, swamps, wetlands, reservoirs and paddy fields. Fishes have played a significant role in social customs and traditions of people in Assam; fish being one of the major food items for about 95% of the population of Assam. Considering the fact that Assam gets a lot of rainfall and fish is an integral part of the diet of the people, the problem of unemployment in the state could also be solved by encouraging entrepreneurship in fish production and trading in fish items. In fact, it has been observed that a good number of unemployed youths are choosing fish culture as a livelihood option these days. There is a lot of demand for local fish of Assam in other parts of the country also because of its unique taste and delicacy. Increased production could also offer more opportunities for the local population to relish better quality fish. One of the reasons of low productivity is the derelict condition of the natural water bodies and lack of initiative of the local populace to use modern tools and techniques of fish culture. The change in the rainfall pattern and lack of adequate irrigation facility in the region are among the major constraints on the development of agriculture and fishery, making the region backward in both the sectors in spite of its inherent potential for sustainable development. The development can be made much more sustainable and the overall socio-economic scenario of the region can also be improved by devising improved means of water resource management.

Water Challenges of the 21st Century

1. Ground water depletion.
2. Degradation of water quality (e.g. arsenic poisoning).
3. Low waste use efficiency.
4. Expensive new water sources (i.e. payment for water consumed).
5. Resource degradation (e.g. pollution, siltation, etc).
6. Health concerns due to water borne diseases.
7. Increasing demand for water in agriculture and industry.

How to solve the present water crisis?

The following steps can be taken :–

(i) Rainfall is sufficient but rainwater needs to be properly harnessed for agriculture, fishery, etc.

(ii) Rainwater harvesting method has to be devised by combing the traditional system with modern technology.

(iii) Sustainable management of water resources by involving the local people in the management.

(iv) Secure water rights to the people and make them aware of the impending water crisis.

(v) Provide appropriate technology for water use and conservation.

Why rainwater harvesting?

Rainwater is the most pure form of water, soft in nature and free from organic matter. It is also the primary source of freshwater and if the rainwater is harnessed it can meet the water demands. The principle of collecting water of strong precipitation from catchments surface is referred to as rainwater harvesting. It consists of the collection of rainwater, storing it for use or for recharging the groundwater, and prevention of its wastage. In the Northeast, there is sufficient rainfall (in spite of the observed declining trend), but it requires appropriate harvesting technology and management.

Advantages of rainwater harvesting

1. Meeting the water demands in areas where rainfall is sufficient but surface water is either not adequate to its needs or not of good quality. In Barak Valley it can be used to ensure adequate supply of water for agriculture during low rainfall season and for improving fish production.
2. Recharging the underground water level in urban areas where water penetration to ground level is poor.
3. Reducing urban flooding.
4. Utilizing the water which would otherwise be lost.
5. Can be stored for irrigation of agricultural land by building water reservoir. For example, in Rajasthan the building of 'Johads' under the initiative of Rajendra Singh has solved the water problem to a large extent. In Barak Valley such initiatives can help cultivation in winter season.
6. Helps the growth of natural vegetation and keeps the climate stable.
7. Improves the quality of groundwater in aquifers and helps conserve soil moisture.

8. It also reduces power consumption, otherwise necessary for obtaining water for irrigation using groundwater. In Barak Valley it will help irrigating fields where irrigation facility is hopelessly poor.

Suggestions

(i) Identification of suitable catchments area for successful rainwater harvesting.

(ii) Selection of appropriate sites.

(iii) Studying topography of the area and estimation of water demand for designing suitable methods for harvesting.

(iv) Integrating traditional water harvesting systems with modern technology for collection and utilization of rainwater.

(v) Encouraging community participation and involvement of local NGOs for the purpose of successful natural water resource management.

(vi) Restoring the derelict natural water bodies to serve as perennial water reservoirs for irrigating agricultural fields and for fish culture.

(vii) Reviving the individual and community ponds and tanks.

(viii) Involving the SHGs and local cooperative bodies in water harvesting and water management.

(ix) Exploring all possible means of water harvesting, conservation and management for sustainable agriculture.

REFERENCES

Das, P.J. (2004), 'Rainfall Regime of Northeast India: A Hydrometeorological Study with Special Emphasis on the Brahmaputra Basin', Unpublished Ph.D. thesis, Gauhati University.

Das. P.J. (2009), 'Water and climate induced vulnerability in northeast India: Concerns for environmental security and sustainability', *WATCH Research Report 1, AARANYAK,* Guwahati, Assam, India. www.aaranyak.org 2009.

Dey, M. (2007), 'Rain water Harvesting' in *Water Harvesting and Utilization in North East India*, Assam University Publication, pp. 39-45.

Gupta, S., Buragohain, K & Bhuyan, J.R. (2008), 'Importance of ponds in integrated water resource management in Barak Valley, Assam, India', *Poll. Res.* 27(3): 587-590.

IPCC (2007a), 'Summary for Policymakers', In: *Climate Change*, The Physical Science Basis. Contribution of Working Group I to the Fourth Assessment Report of the Intergovernmental Panel on Climate Change [Solomon, S., D. Qin, M. Manning, Z. Chen, M. Marquis, K.B. Averyt, M.Tignor and H.L. Miller (eds.)]. Cambridge University Press, Cambridge, United Kingdom and New York, NY, USA. .

Maddison, A. (1970), *Economic Progress and Policy in Developing Countries*, George Allen and Unwin Ltd., London.

Mirza, M.M.Q., R.A. Warrick, N.J. Ericksen, & G.J. Kenny (1998), 'Trends and persistence in precipitation in the Ganges, Brahmaputra and Meghna river basins', *Hydrological Sciences-Journal-des Hydrologiques*, Vol. 43(6), pp. 845-858.

Reports of Dept.of Agriculture - *agricuture assam report.*

TOI, March 12th , 2010, "33% water in India undrinkable".

9

Problem of Drinking Water in Cachar District, Assam

Pranay Jyoti Goswami

[**Abstract:** Cachar District, which accommodates the largest population among the districts of South Assam, is one of the educationally and culturally most advanced districts in the state of Assam and the north-eastern region of India. Silchar, the district headquarters, legitimately takes pride as the city where five of the seven states in Northeast meet within thirty to sixty kilometres from the urban ends. It is one of the first three municipalities in Northeast India and the second largest urban conglomeration in the region. One Central University, one National Institute of Technology, one medical college, one polytechnic, three post-graduate training colleges and six degree colleges, besides corresponding numbers of feeder higher secondary and secondary schools, are located in Silchar town which is also endowed with local stations of All India Radio and Doordarshan. Flanked by the north-eastern mountain ranges, the district borders Bangladesh on its west. The mighty river Barak passes through the heart of the district where it is joined by large number of major tributaries, besides streams and rivulets, on both banks, coming down from the neighbouring hills. Large number of natural water bodies – wetlands, lakes and oxbows – within the district is amongst its precious water resources. Flood is a perennial problem, which visits the district almost every year. On the other hand, the scarcity of drinking water is the cry of the people of the district. The paper analyzes the problem and suggests viable measures for its sustainable resolution.]

Introduction

In India, the demand for water is increasing continuously both in urban and rural areas. This is largely due to socio-economic development and population growth. There are also tensions and disputes over sharing and control of water resources. The emerging scarcity of water has also raised a host of issues related to sustainability of the present form of economic development, sustained water supply, equity and social justice, water financing, pricing, governance and management. The problem is indeed global and coupled with other dimensions. It is indicated in chapter 18 of Agenda 21 of the UN Conference on Environment and Development (UNCED) that about "eighty per cent of all diseases and over one third of deaths in developing countries are caused by the consumption of contaminated water, and on an average as much as one- tenth of each person's productive time is sacrificed to water related diseases." It is reported by the Government of India in the Parliament that in India (i) the iron content in groundwater are higher than those prescribed in 254 districts, (ii) the fluoride levels have breached the safe level in 224 districts, (iii) the salinity had risen beyond tolerance levels in 162 districts, and (iv) the arsenic levels were far more than permissible limits in 34 districts (*The Times of India*, March 13, 2010). The people have also accelerated the crisis in India by increasing population and aspirations.

The cbjectives of the present paper are:

1) To give an idea of the existing status of supply of drinking water in the rural areas of Cachar district.

2) To provide an overview of the existing status of supply of drinking water in the urban areas of Cachar district.

3) To suggest some remedial measures to solve the problem of drinking water in the Cachar district.

The study is based on both primary and secondary data. The primary data have been gathered by the author from the field in the Barak Valley, while the secondary are those collected from the office of the Silchar Municipal Board, office of the Executive Engineer, PHED, Cachar, Silchar, offices of PHED, Div-I & Div-II, and office of the Assam Urban Water Supply and Sewerage Board at Silchar. For the sake of analysis, the study is divided into four sections : Section I covers an overview of Cachar District; Section II deals with rural water supply in Cachar District; Urban water supply is discussed in Section III; and Section IV presents conclusions and suggestions.

SECTION - I

A Brief Profile of Cachar District

As per 2001 census, the total geographical area of Cachar district of Assam is 3,786 sq. km out of which the rural area covers 3,751.37 sq. km. The district is divided into two subdivisions, viz. Silchar and Lakhipur. There are 15 blocks with 1,047 villages in the district. Out of 1,047 villages, 1,020 are inhabited. As per the same census, the total population of the district was 14,44,921, out of which the number of rural population was 12,43,534. Thus 86.6 percent of the total population in the

Table 1: Cachar District at a Glance

Particulars	*Total*	*Rural*	*Urban*
Area (Square K.M.)	3786	3751.37	34.63
The number of households	273694	234326	39368
Population	1444921	1243534	201387
0-6 Population	229219	206898	22321
Literates	824437	671403 (64.77%)	153034 (85.46%)
Total workers	465734	403310	62424
Cultivators	107305	106892	413
Agricultural Workers	62576	62398	178
House hold industry workers	14877	14074	803
Other workers	280976	219946	6130

Source: *Assam at a glance*, Data Dissemination Cell, Directorate of Census Operations, Assam. and Statistical Hand Book , Assam, 2005.

district live in rural areas. 64.77 percent of the rural population of the district is literate as against 85.46 percent in the urban areas. The total number of male workers in the rural areas in the district was 3,07,397, out of which 89,138 were cultivators and 29,810 belonged to agricultural labourers. 38.68 percent of total main workers depended on agriculture. Table I shows the statistical data related to Cachar District.

The climate of the district is marked by excessive humidity. The hot and cold months are very distinct. The average number of rainy days in a year ranges in between 100 to 140 days. .The average annual rainfall of the district approximately ranges in between 2000 to 3550 mm. November to March is the dry (winter) season in the district where monthly rainfall ranges in between 3.0 mm to 24.0 mm. The average monthly rainfall in Cachar district in the year 2006 is given in Table-2.

Table 2 : Monthly Rainfall in Cachar District (in mm)

Month	*Jan*	*Feb*	*Mar*	*Apr*	*May*	*Jun*	*Jul*	*Aug*	*Sep*	*Oct*	*Nov*	*Dec*	*Annual*
Rainfall	0	50.4	3.5	151.3	260.6	660.7	498.1	230.9	150.0	48.9	25.0	0.4	2086.8

Source : *Statistical Handbook Assam*, 2007.

SECTION II

Before the introduction of National Water Supply and Sanitation Programme in social welfare sector in 1954, the provision of drinking water in the rural areas was the sole responsibility of the states. The states at that time first set up Public Health Engineering Departments (PHEDs) to attend to the service. Despite the efforts made by the state and central governments, it was found during mid-1960s that rural water supply schemes could not reach the far-flung interior villages. To accelerate the full coverage of the problem villages, the Government of India introduced in 1973 the Accelerated Rural Water Supply Programme (ARWSP) Scheme with a view to assisting the States and Union Territories with 100 percent grant-in-aid for implementing the schemes in such villages. This programmes continued till 1973-74. But with the introduction of Minimum Needs Programmes

(MNP) in the 5th plan, it was withdrawn in 1974-75, but it was reintroduced in 1977-78 finding that the progress of supply of safe drinking water to identified problem villages was not as per expectations.

In 1985 the subject of rural water supply was transferred from the Ministry of Urban Development to the Department of Rural Development. In 1986, the National Drinking Water Mission (renamed as Rajiv Gandhi National Drinking Water Mission in 1991) was set up. It is through the mission and ARWSP that the Government of India has been trying to attend the problem of rural water supply. The prime objective of the Mission was to provide at least one source of drinking water in a village during a decade; and subsequently, it focused on the following major objectives:

1. To ensure coverage of all rural habitations especially to reach the unreached with access to safe drinking water.
2. To ensure the sustainability of the system and sources.
3. To tackle the problem of water quality in affected habitations and to preserve quality of water by institutionalising water quality monitoring and surveillance through a catchments area approach.
4. To adopt community participation in rural water supply programme through reform project in the country on a pilot basis.

In this context, Sector Reform Programme introduced in 1999 is an improvement as it included community participation, awareness creation, community contribution and rural transformation inclusive of operation and maintenance. In the same year, the Department of Drinking Water Supply was set up within the Ministry of Rural Development to extend greater thrust to the water and sanitation programmes. In drinking water supply, Sector Reform Project (SRP) was launched in 67 districts on pilot basis, which were later on scaled up in 2002 across the country in the form of Sajal-Dhara under its comprehensive guidelines. It aimed at providing operational flexibility to the state governments and implementation flexibility to the districts and to the Gaon Panchayet level institutions. In 2003, the Government of India made a survey to know the status of habitations, i.e. whether the habitation is fully covered (FC), partially covered (PC) or not covered (NC). Status of habitations in India, Assam and Cachar as per 2003 survey is shown in Table-3.

Table-3 reveals that the percentage of Not Covered (NC) villages in Assam (36.48) is higher than that of all India percentage (16.9). Again, the percentage of NC habitations in Cachar District (50.1) is higher than that of Assam as a whole (36.48). As per National Habitation Survey 2003, there are many habitations, water in which is affected by Chemical contamination. In 2008, in the website of the Ministry of Rural Development, it was seen that in Assam water of 15,567 habitations are iron contaminated, 6 arsenic affected, 85 fluoride affected and 11 nitrate affected. The problem of Chemical contaminations in some habitations of Cachar district will be discussed later on. The block-wise status of habitations in Cachar as per 2003 survey is shown in Table-4.

Table 3 : Status of habitations in India, Assam and Cachar (2003 survey)

	No of habitations All India	*Percent*	*No of habitations All Assam*	*Percent*	*No of habitations All Cachar*	*Percent*
Total	1599433	100	80468	100	2562	100
Fully covered (FC)	916382	57.2	27300	33.92	506	19.7
Partially covered (PC)	412646	25.7	23813	29.59	773	30.2
Not covered (NC)	270405	16.9	29355	36.48	1283	50.1

Source: National Habitation Survey, 2003.

Table 4 : Block wise status of Habitations

Name of Block	*Number of FC habitations*	*Number of PC habitations*	*Number of NC habitations*	*Total* Habitations
Borjalenga	30	39	89 (56.3)	158
Udharbond	122	124	39 (13.6)	285
Borkhola	24	82	136 (56.1)	242
Baskandi	27	25	59 (53.1)	111
Binnakandi	26	39	38 (36.8)	103
Kalain	27	63	146 (61.8)	236
Katigorah	07	30	60 (61.8)	97
Lakhipur	43	60	27 (20.7)	130
Narsingpur	34	45	221 (73.6)	300
Palanghat	20	30	171 (77.3)	221
Rajabazar	78	61	38 (21.4)	177
Salchapra	18	38	20 (26.3)	76
Silchar	20	66	34 (28.3)	120
Sonai	28	47	163 (68.4)	238
Tapang	02	24	42 (61.7)	68
Total	506	773	1283	2562

Source: *National Habitation Survey*, 2003.

Note: The figures in the brackets indicate the percentage of total habitations.

Table-4 reveals that 6 blocks are having the problem of drinking water as more than 60 percent of the habitations in those blocks are not covered by drinking water facilities.

Management of Drinking Water

Public Health Engineering Department (PHED) is the authority for the supply of drinking water in the habitations. There are two divisions of PHED in Cachar District – Division 1 and Division 2. The census report 2001 shows that in all inhabited villages there are sources of drinking water. However, there are many habitations in a village which are not yet covered under the scheme as per the habitation survey, 2003.

The main source of supply of water in Cachar District is the river Barak. The other sources are the tributaries of river Barak, ring-wells and tube-wells. Different schemes in operations in the district are shown in Table-5.

Table 5: Schemes of PHED, Cachar District

Division	*Ring well*	*Tara pump and other Pump*	Pipe water supply scheme
I	750	800	121
II	2005	2898	188

Source : PHE Offices, Div-I and Div-II

There were 18 rainwater harvesting schemes in the district in 1997, but these schemes are now defunct. The main sources of pipe water schemes are rivers and deep tube-wells, as shown in Table-6.

Table 6: Sources of Water in Pipe Water Schemes in Cachar

Division	*Deep tube-well*	*River*	*Total*
I	3	118	121
II	35	153	188
Total	38	271	309

There are many habitations in Cachar district which has been detected as arsenic affected. As per official records of Silchar Division I, as many as 53 habitations have been detected as arsenic affected and 170 habitations are iron affected. It has also been learnt from Chief Engineer, PHED's letter no. PHE-89\PLG\2005-06\16-21 dated 4-4-06, that there are 5 arsenic villages in the Silchar Division II. These are Pangram (Udarbond), Durganagar part V (Udarbond), Satkhorakandi Part II (Sonai), Kamranga village (Rajabazar) and Modi Basti (Rajabazar). There are also many habitations which are affected by Chemical contamination. For example, in the Kalain Block, out of 236 habitations 29 are affected by Chemical contaminations. Recently the problems of chemical contamination in the identified habitations are being solved by introducing pipe water scheme from the alternative sources of water.

In this context it may be useful to know that the following norms have been adopted for providing safe drinking water to rural population in the country which are accepted by the Government of Assam also:

a) 40 litres of safe drinking water for per capita per day (lpcd)

b) 30 lpcd additional for cattle in the Desert Development Programme Areas. [Not applied for Cachar district]

c) One hand pump or stand post for every 250 persons.

d) The water schemes should exist within 1.6 km in the plains and within 100 meter elevation in the hilly areas.

The quality of service rendered by PHED where there are water supply schemes is another area that deserves to be focused. The findings of the survey conducted by Community Polytechnic Cell, Silchar Polytechnic, Silchar in 1994 in 312 habitations situated in 12 blocks of Cachar district revealed that "Many of the installed piped water supply schemes.... are not functioning due to lack of maintenance by the PHED. The PHED could not provide recurring expenditure to run the plants. The initial expenditure of construction of water supply schemes are usually borne by the government of India but the responsibility of the maintenance rest with State Government through the PHED. But it has been found during the field survey that paucity of funds stand in the way of proper maintenance of these units. Unless this problem is sorted out, it is feared that the very basic purpose of such schemes would be defeated." As per the report of the Validation Survey conducted by Community Polytechnic, Silchar, "installation of tube well, Mark II tube well, ring well etc are not successful in Cachar district as the iron content in ground water is more than 7ppm in some areas." As regard supply of pipe water, because of the delay of repairing cracked pipes and also because of lack of proper chlorination the people are not getting safe drinking water. Uneven flow of water, unscientific selection of street water hydrants is responsible for inequitable distribution of water. These problems still exist in the district.

Distribution of drinking water in the villages through pipelines

According to the report of a group of experts directed by Michel Camdessus, former Director-General of the IMF, the financing of the water must obey the principle "you use it, You pay for it" and through recourse to the capital market. This principle is also adopted by Government of India by institutionalising the scheme (tape water supply). The village panchayats are given authority to manage the scheme. The Government of Assam in 1995 prepared an Act for distribution of pipe water by constituting a Drinking and Sanitation Society. The society is to maintain the cost of supply of water. As per the initial guidelines, house connection facilities are to be given to the applicants provided that he or she must pay Rs. 600/- as non-refundable deposit for household level and Rs. 1500/- for hotel and shop. The monthly fee of Rs. 50/- was fixed for households and Rs. 100/- for hotels and shops. The society was given power to charge money even from the person who collects water from street hydrants. These rates can be changed as and when maintenance cost changes. The government has given power to the PHED to take over the charge of the society when the society fails to manage the affairs as per the guidelines issued by the government. It is learnt that some societies have already been constituted in the districts but majority of them are not functioning properly.

SECTION III

A number of factors like climate, culture, food habits, work and working condition levels and type of developments and physiology determine the requirement of water. Regarding the minimum per capita per day litre of water, urban authorities of different cities/ towns have fixed different norms. As per the Bureau of Indian Standards, IS:1172-1993, a minimum water supply of 200 litres per capita per day (lpcd) should be provided for domestic consumptions with full flushing systems. The National Commission on Urbanisation (1988) recommended that a per capita water supply of 90- 100 litres per day is needed to lead a hygienic existence, and emphasised that this level of water supply must be ensured to all citizens. In Greater Mumbai 135 lpcd is fixed as the minimum domestic requirement of water, but the Delhi Development Authority considers 225 lpcd and additional 75 lpcd necessary for industrial, commercial and civic uses, thus making the total requirement of 300 lpcd.

There are two towns in Cachar District, viz. Silchar and Lakhipur, which are managed by Municipal Board and Town Committee respectively. The statistical data related to Silchar and Lakhipur are given in Table –7.

Table 7: Statistical data related to Silchar & Lakhipur

Town	*Number of Household*	*Total Population*	*Literacy Rate*	*Percentage of SC*	*percentage of ST*
Silchar (MB+OG)	31088	156948	86.37	10.39	.28
Lakhipur	1733	9802	84.49	2.73	.57

Source: Statistical Handbook Assam, 2007.

In Silchar town pipe water supply scheme was started in the year 1971, monitored by the Silchar Municipal Board. The capacity of the plant was 5 MGD but recently it has been increased to 6 MGD. There were 40,001 house connections in Silchar municipality area in February 2010 and the number of street hydrant was 500. The monthly charge of the municipality for water supply is Rs 70 only. But for slum areas, the charge is Rs 95 because people of slum areas are exempted from paying other taxes charged by the municipality.

The Government of Assam, in order to provide potable water, enacted the Assam Urban Water Supply and Sewerage Board Act, 1985. The foundation stone of one of the offices was laid at Silchar on 22nd April, 1999 and the plant was inaugurated on 25th February 2006 to cover the wards 13(p), 15(p), 17(p), 18(p) and 16 (full). The capacity of the project is 8.1 MLD and the capacity of the pump is 450 cubic metres. The total number of house connections given by the Board is 1800. There is no provision for street hydrants. Use and pay is the logic behind the fixation of the price of the water. The present rates charged by the Board is given below:-

1000 Litres to 30,000 Litres = Rs. 18 per thousand Litres

30,001 Litres to 50, 000 Litres = Rs. 20 per thousand Litres

Above 50,000 Litres = Rs. 22 per thousand Litres

In Lakhipur also pipe water is supplied to the houses. The river Barak is the main source of water.

Section IV

Conclusions

1. The percentage of Not Covered (NC) villages in Assam (36.48) is higher than that of all India percentage (16.9). Again, the percentage of NC Habitations in Cachar District (50.1) is higher than that of Assam (36.48).

2. As many as 6 Blocks of Cachar are having the severe problem of drinking water, as more than sixty percent of the habitations are not covered by drinking water facilities.

3. The maximum rainfall occurs in the districts mainly during the months of April to August. The surplus water during this rainy season is drained out to Bangladesh. The valley also witnesses severe floods almost every year. So, the sustainable water management should also include solution of flood.

4. In the urban areas about 70 percent households are getting sufficient water. The PHE plant of Sadarghat (Silchar) treats 200 lakh litres water daily. But due to distributional failure the households of certain areas are facing the problem of insufficient water.

5. The plant can not be properly utilised due to load shedding problems. Regular power supply should be ensured in order to supply potable water regularly and sufficiently.

6. The government is gradually withdrawing its responsibility of providing free drinking water.

Suggestions for Solution

1) The highest priority should be given to ensure coverage of remaining PC and NC habitations with sustainable supply of safe drinking water
2) Quality affected habitations should be covered with safe drinking water on a sustainable basis.
3) The SC/ST population and BPL families to be given priority while ensuring coverage of NC/PC habitations.
4) People in the hill areas can be educated regarding the usefulness of rainwater harvesting.
5) Provision should be made for supply of safe drinking water in the school, colleges and other institutions.
6) From the welfare point of view, right of drinking water should be ensured. The government cannot withdraw subsidy element totally. The supply of water through street hydrants should be free from charges.
7) More rural water quality testing laboratory should be set up. In Cachar there are two testing centres; one at Barjalenga and another at Silchar. The first one is for arsenic detection and other for measuring the content of iron.
8) Measures should be taken to organise awareness programmes for collection, storage, handling and consumption of water.
9) Measures should be taken to clean up the polluted rivers, lakes and ponds.
10) Scientists should take up studies for development of low cost water saving/ purification technology as a major challenge.
11) Water quality monitoring and surveillance system should be strengthened.

REFERENCES

Bohra, P.H. (2001), 'Water Harvesting: a Suggestion', *Yojana*, July 2001.

Chand, Dinesh (2003), 'Towards Sustainable Rural Water Supply', *Kurukshetra*, October 2003.

Chand, Dinesh (2002), 'Drinking Water Supply and Availability in rural areas' *Kurukshetra*, October 2002.

Duggal, V.K. (2005), 'Drinking Water and Sanitation', *Yojona*, June 2005.

Durgaprasad, P. (2003), 'Drinking Water; Management Strategies and Future Agenda', *Kurukshetra*, October 2003, p. 10.

Goswami, Pranay Jyoti (2006), 'Status of Potable Water Supply in the Rural Areas of North East India with Special Reference to Barak Valley', in Datta Ray, B. and R.P. Athparia (ed.), *Water and Water Resource Management*, Omsons Publications, New Delhi.

National Habitation Survey, 2003.

Nautiyal, Suresh (2002), 'Water Crisis Agenda for the Century', *Yojana*, February 2002.

Petrella, Recardo (2005), 'Poverty, Water and Globilisation', *Yojana*, June 2005, p. 18.

PHED, Govt. of Assam (1995), *Gramyo Pani-jogan achonir porichalona swamitiupobidhi- 1995.*(Guidelines for managing committee of rural water supply scheme,1995), Guwahati.

Ponnuraj, S. (2003), 'Role of Panchayti Raj Institution in Drinking Water Supply Scheme', *Kurukshetra*, October 2003.

Rao, M.J. Mohan (2003), 'Community Intervention in Traditional Rain Water Harvesting- Drinking Water', NIRD, Hyderabad, November 2003.

Singh, K.P. (1996), 'Water Harvesting', *Yojana*, November 1996.

Swaminathan, D. (1995), 'Ensuring Supply of Safe Drinking Water', *Yojana*, January 26, 1995, p. 17.

The Times of India, March 13, 2010.

10

Water Resource Potential of the Pakai Beel, Cachar, Assam

Bimal Sinha

[**Abstract:** Water is one of the important natural resources available to the humankind. The life depends on water and throughout the ages the people have been living on water. The volume of water in the planet is limited. Only 2.5% of the world water is available for use and two-third of this is locked in the form of ice caps and glaciers. We use 70% of the water in agriculture. India has 2% of land in the world, 4% of fresh water, 16% of population and 10% of cattles. The annual rainfall in India is approximately 4000 billion cubic meters. The water potential of the country is estimated at 1869 billion cubic meters. The Barak Valley, in India's Northeast, is surrounded by the hills on three sides, and the average rainfall in the valley is 2500-3000 mm. The main river, Barak is connected to innumerable streams and rivulets which originate in the neighboring hills. The water-logging is the main cause of floods which inundate the low lying areas in the valley during the monsoons. The people have been facing remorseful and disastrous flood damages since time immemorial, but the frequency and intensity of floods have increased over the years. The irregular and heavy rainfall in the summer season, water logging, silting in the rivers and rivulets, land erosion, etc. are the main causes of floods and inundation of the low lying areas and human habitations. The loss of life and property and the damages to standing crops and cattles are the calamities that the people in this valley have to face almost every year. The people in the villages are marooned during the floods, and many die of hunger and diseases without treatment. On the other hand, there is scarcity of water for drinking and domestic usages or for supply to the paddy fields. The situation demands sustainable development of wetlands or water bodies in the valley, locally called *beel* and *haor*. The Pakai Beel, situated beside the NH-54, under Narsingpur GP, in the Cachar district of Assam, is one such important water body which deserves attention for sustainable development. It is connected with the Chatla Haor by a rivulet and it is considered as one of the important natural water reservoirs in Barak Valley. The key words used are beel, haor, water resource, water reservoir, water logging, inundation, scarcity, effluent, rivulet.]

Introduction

Water is one of the most essential items needed by human, plants and other living beings for their survival. Water also maintains the ecological balance between various groups of living organisms and the environment. Water is the most precious resource on earth, covering 70% of land. The ecological degradation in the poor and developing countries has assumed emergency proportions largely because of the mindless commitment of the capitalist path of development to the economic growth strategy. The costs of such development are manifested not only in terms of class conflicts and economic exploitation of the weaker sections by the rich and the powerful but also in the reduction of the natural resource base on which the poor depend for their livelihood. The environmental problems are usually found within a dependency framework in which economic development model provides the parameter. (Sivakami: 2002)

The aquatic ecosystems are extremely important to the humankind as they have various uses, including drinking water, irrigation for agriculture, navigation, recreation, etc. and these are also the sources of organic productivity. The water bodies serve as cheap receptacles for all kinds of waste disposal. The freshwater ecosystem can be broadly divided into two habitats – lentic and lotic. Lentic habitats are the standing water bodies including lakes, reservoirs, ponds, pools and swamps in which the water is without any flow in a definite direction, while the lotic habitats are the running water systems like rivers and streams in which water is continuously flowing in particular direction (Kaushik & Saksena, 1999). The Pakai Beel comes under natural standing water reservoir category.

Area of the Study

The Pakai Beel is situated within the jurisdiction of the Narsingpur Development Block (DB) and Narsingpur Gaon Panchayat (GP). The total geographical area of the Block is 9606 hectares. The river Rukni, a tributary of the great river Barak, passes through the Narsingpur GP area. There are also some small streams or rivulets within the jurisdiction of the same GP. Bekirpar Anua is another such natural standing water reservoir in the area, while the great fen of the Barak Valley – Chatla Haor-- is also not far away. The *beels* and *haors* in the area are interconnected by the rivulets. The village in which Pakai Beel is situated is called Pakaipar. The total area of the Pakai Beel is 120.1 hectares according to the Second Re-Survey (2005). It is 20 kms from Silchar town (headquarters of Cachar district) and ½ km from Kabuganj Bazar. Kabuganj is an important market township (*bazar*) in the area.

Objectives of the Study

To determine the possibility of water harvesting and its proper utilization in the area surrounding the Pakai Beel; to assess the commercial prospects of water resources of the area; to evolve the socio-economic status of the farmers and other categories of local people; and to outline the problems and prospects of socio-economic development of the area.

Relevance of the Study

Water is one of the most vital resources for all kinds of life on earth. Human had been closely associated with water ever since it originated as a specie. In a developing country like India, 70% of the population is dependant on agriculture. The regular supply of water is crucial for the development of agriculture. The cultivators in the country do depend mainly on the monsoon for

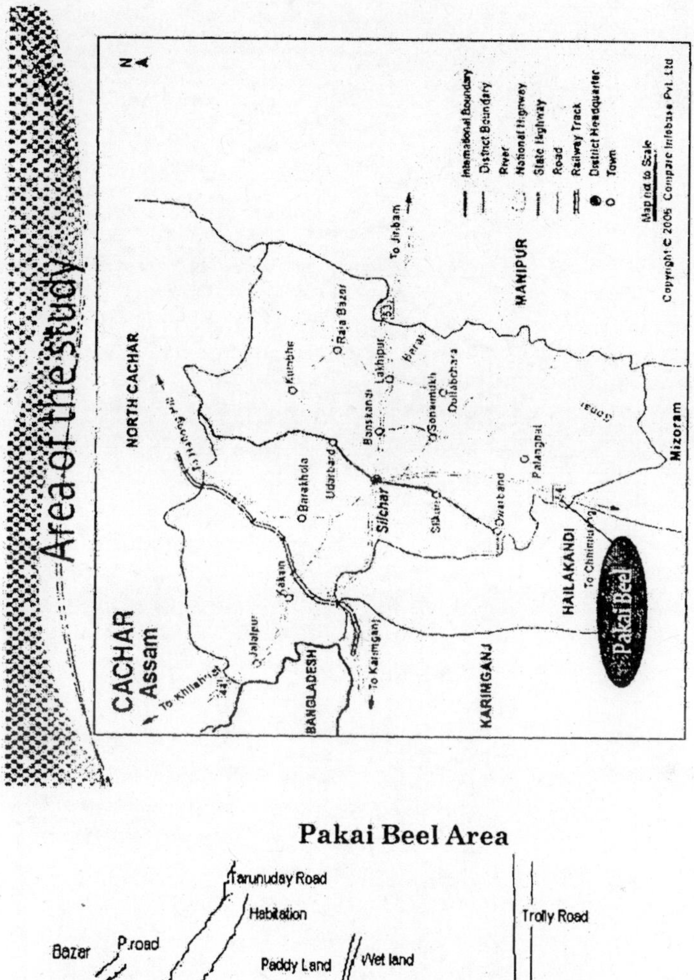

Pakai Beel Area

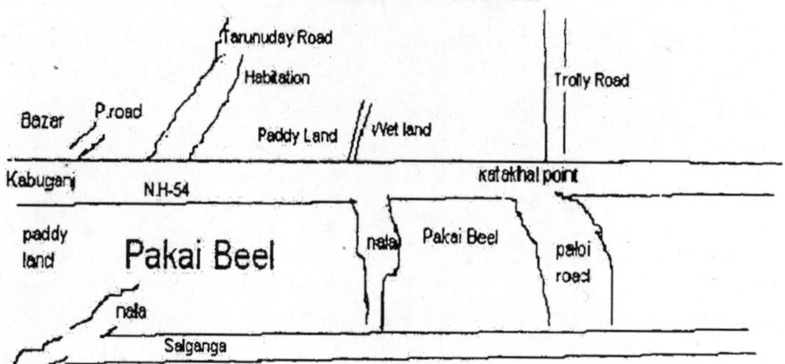

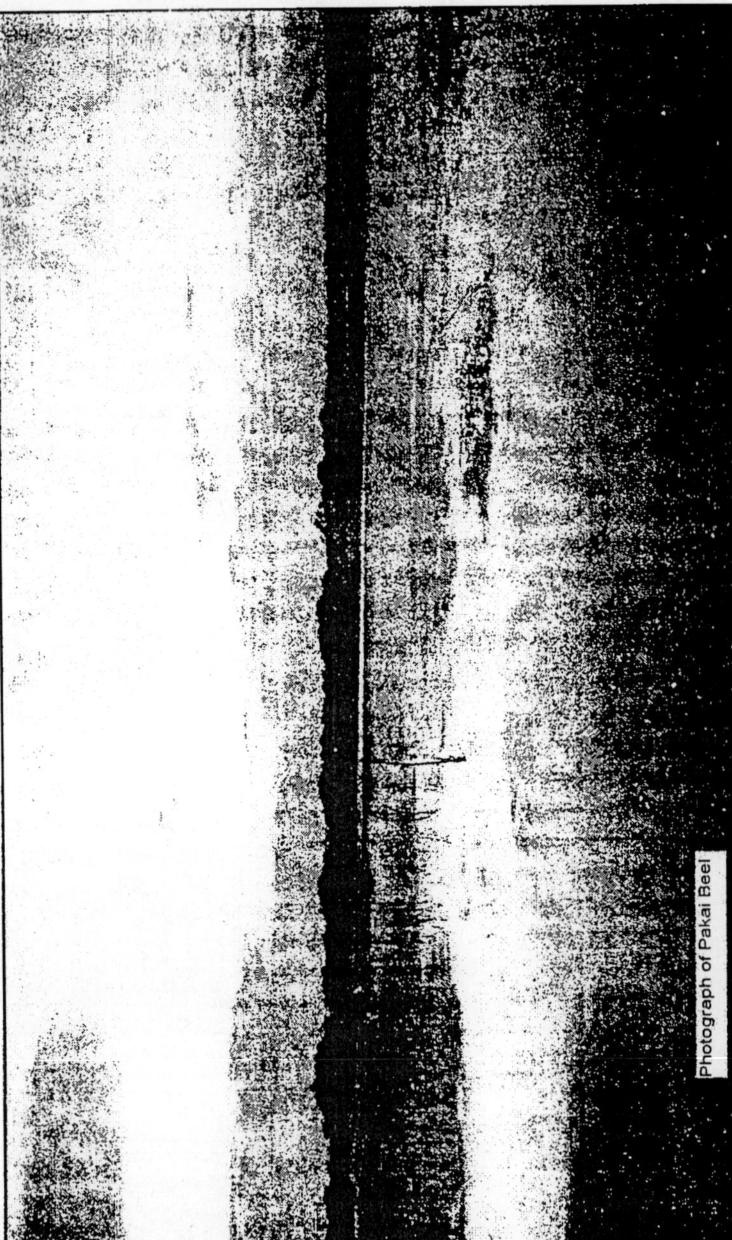

Photograph of Pakai Beel

the supply of water to the agricultural fields. Several regions in the country face drought in the agricultural season and a large number of farmers commit suicide every year. In such circumstances the development and maintenance of the natural water bodies and water harvesting should be considered as a key collective responsibility of the state and all citizens. The study, therefore, included agriculture, water harvesting, public health, commercial application of water, peoples' participation, human resource development, and the government's schemes for development and management of the wetlands as relevant areas in context of the sustainable development of the Pakai Beel.

(i) **Agriculture:** The mainstay of economy in the Barak Valley region of Assam is agriculture. In the Narsingpur GP, the total number of the farmer families is 1275, the total crop area is 717.5 hectares, and the quality of soil is sandy loam. The cropping status is paddy and seasonal crops. The area under *Aus/Ahu* paddy is 94 hectares, *Sali* paddy 638 hectares, summer pulses 0.5 hectares and the *Rabi* crops 93 hectares. In the Pakai Beel, *Boro* paddy and early *Ahu* can be cultivated, while the dyke areas can be suitable for horticulture of different varieties. In addition to agriculture and horticulture, the Pakai Beel can boost fish farming and livestock development under proper incentives and management.

Table 1: Agricultural Cropping Scenario of Narsingpur GP

Kharif Crop	*Crop Area (ha)*	*Rabi Crop*		*Farm Machineries*		*Total Area*
		Name	Area (ha)	Equipments	No.items	770 ha
Early Ahu	Nil	Black green	12	Power Tiller	12 nos	Net crop
Boro Paddy	Nil	Mug	1	Power pump	14 nos	717.5 (ha)
Regular Ahu	94	French Bean	18	Sprayer	15 nos	
Sali	638	Oil seed (M)	22			
Summer pulse	0.5	Summer veg.	13			
		Winter veg.	21			
Total Kharif crop	732.5	Total Rabicrop	93	Total	42	

(ii) **Water harvesting:** The rain water can be stored in the catchments area by making dykes and embankments, digging canals and removing the silts from the rivulets and canals. In the Pakai Beel area embankment expenditure will not be heavy because two sides of the *beel* already have good dykes. The rainwater can be harvested under proper watershed management programme. Narsingpur is situated on the bank of the river Rukni which originates in Mizoram. In this area there is another natural water reservoir, called Bekirpar Anua, which is connected with river Rukni. At present, Irrigation Department, Govt. of Assam, is making a canal from Bekirpar Anua to Pakai Beel. Thus, the Narsingpur GP has both stream water and stable water resources which should be utilized through a proper strategy for the sustainable development of the area.

(iii) **Public Health:** The scarcity of pure drinking water is one of the major causes of the recurrence of epidemic diseases and ill health of the rural people in Barak Valley. In the Cachar district the survey of drinking water facility in 2003-04 in a sample area of 75 habitations revealed that 74 out of 75 habitats were only partially covered and one habitation was not at all covered by the scheme (*Statistical Handbook, Assam, 2005*). The availability of drinking water is also an indicator of the overall water management in an area. It is suggested that the exploration of the water potential of Pakai Beel can be useful not only for the inhabitants of the Narshingpur area but also for the livestock and other living organisms which are essential for maintaining the ecological balance.

Table 2 : Water Resources available in the Narsingpur GP Area

Ponds	*Seasonally inundated floodplain*	*Swamp/ Marsh*	*Ox-bow lake*	*Total land area*
1100 nos.	7 hectares	130 hectares	5 hectares	770 hectares

(iv) **Commercial prospects :** In parts of the Pakai Beel the land is suitable for the cultivation of *Boro* and early *Ahu* paddy. The local people also practice fishing, hatchery, and duckery in the *beel*. A good number of fishermen maintain their families mainly out of the income from fishing in Pakai Beel. It has been suggested by many that vermin-compose can be locally manufactured in the area by using water hyacinths and weeds etc of the *beel*. Horticultural crops, like lemon, coconut, black peeper, banana, mango, guava, orange, zinger and turmeric, can be grown in the dykes and embankments of the *beel* as well as in the areas surrounding the *beel*, provided those are developed and maintained properly. Above all, the water reserved in the *beel* can be used for agricultural and horticultural activities and for domestic use in the surrounding areas of the *beel*. The development of Pakai Beel is thus associated with the overall economic development of a large area under the Narsingpur GP.

(v) **Peoples' participation:** The historical experience of the development initiatives in the Barak Valley over a long period of time shows that no initiative is taken by the government without popular demand on the part of the people and no initiative can be successful after it has been taken up without the support and cooperation of the people. There are also certain schemes for rural development which cannot be successful any where without the people's participation. The sustainable development of the water bodies is one such area that requires awareness and active participation of the local community. It is, therefore, proposed that the local people in the Narsingpur area should be motivated to develop the Pakai Beel, to maintain the dykes and canals and to keep the water body clean. They should be involved in decision making and development works not only related to the *beel* but also any other development initiative in the area. In fact, in the Narsingpur area there are already 12 SHGs and 4 NGOs actively involved in various developmental services to the people. These voluntary organizations have tremendous influence in the area and they are capable of uniting and mobilizing the local people. The scheme for sustainable development of the Pakai Beel may, therefore, be evolved by forming a SHG or NGO with the objective of socio-economic improvement of the stakeholders along with the development and maintenance of the water body. The government alone cannot deliver all that are needed by the people, even if it has the will to do so. Therefore, the people's participation becomes important in all sustainable development initiatives.

(vi) Human Resource Development : The farmers in the Narsingpur area are generally the seasonal workers as they have to depend only on the monsoon for supply of water to their agricultural fields. The income generated through seasonal employment can hardly be sufficient to meet the minimum needs of the families throughout the year. The people in the area are, therefore, generally poor. On the other hand, it is observed that with proper water harvesting and watershed management programme they can be engaged in cultivation of a variety of crops in the same fields throughout the year. Land in the Narsingpur area is highly fertile. Narsingpur is also one of the highly well connected areas in the Barak Valley. The NH-54 or Aizwal Road is adjacent to Pakai Beel. It is only 20 Kms from the Silchar town (headquarters of Cachar district) and ½ km from Kabuganj Bazar which is one of the important business centres in the valley. If Pakai Beel can be developed into a major fishery, for which it has all the potential, the wholesale marketing of fish would not be a problem. The agricultural and horticultural products of the area also can be marketed in the Kabuganj Bazar. With proper agricultural, horticultural and livestock development, the unemployed local youths in the area can be motivated to specialize in various occupations and professions, including business and other small scale enterprises. The people in the area can be motivated to self-employment schemes and entrepreneurship can grow locally. The economic reconstruction programme in the Narsingpur GP can be initiated with Pakai Beel as the centre of rural development.

(vii) **Government Schemes**: There are some schemes of the government for the development and maintenance of wetlands and water bodies which are operated through block development authorities. The NREGS schemes currently in operation under Narsingpur Block of Narsingpur Goan Panchayat (GP) are the following:

1) Cleaning and desilting of Jumjumi Khal and Salganga Nala - Rs 4.99, 600 (2008-09).

2) Desilting and excavation of a *nala* (canal/drainage) from village Katakhal to river Salganga - Rs. 3, 15, 000 (2007-08).

3) Desilting and excavation of irrigation *nala* for agricultural purposes and Ipomea cleaning in the Pakai Beel, Narsingpur pt. IV - Rs. 5.74, 000 (2007-08).

4) An irrigation scheme under the Department of Irrigation, Govt. of Assam, from Bekirpar Anua – Rs. 72,34,714 (2008-09).

The schemes are in operation and the initiative of the government in this matter is generally appreciated by the people. However, the schemes are managed officially and there is hardly any scope for the people to participate in the activities. The execution is very slow and the utilization of funds is not known to the people.

Towards Problem Resolution

The Barak Valley is an economically backward region within the state of Assam. Over the decades the development of the Narsingpur area has been neglected on the plea of financial constraints. The lack of awareness among the local people of the potential of the wetlands and water bodies has also systematically contributed to the deterioration of those bountiful gifts of nature to the people of the area. In the process, the natural water bodies have been converted into dustbins, hunting grounds of mosquitoes and insects and the major sources of epidemic diseases. The Pakai Beel today is covered by ipomeas and other weeds and the natural drainages are filled by continuous silting. Despite heavy rainfall in the catchments area, water cannot be preserved in these natural reservoirs properly. Flash floods and water-logging problems are also due to the deteriorating conditions of those natural reservoirs. The inundation of the low lying paddy lands, fisheries, hatcheries and human habitats are very common. On the other hand, these problems are because of the indifferent and callous attitude of the civil society and the administration. The over-flooding and water-logging are basically due to the declining capacity of the natural water bodies to harvest and retain the seasonal flow of excess water. The water resource is thus wasted without utilization because of the absence of awareness among the people and proper planning on the part of the authority. The need of the hour is sensitizing the people about their role in the sustainable development of natural resources of the area. The people have to be motivated to participate in the development processes as legitimate stakeholders. The group farming and SHG activism are important for sustainable development and commercial application of the resources in the area without affecting the ecosystems and natural environment. The development has to be sustainable, keeping in view the interests of the future generations and the needs of the present to generate income for livelihood. The macro-economic approach for the development of the Narsingpur GP area should take into account those facts and fully exploit the potential of the natural resources, including the water bodies, in a sustainable manner. The government of the state and at the centre should play the respective assigned roles to end the prolonged era of indifference and neglect to this part of Assam. The remedial measures initiated by the government through various schemes for micro-economic development in various parts of the country should reach Narisingpur GP without further loss of time. The sustainable development of the Pakai Beel can be identified as a pilot project to make a new beginning which can eventually impact the entire valley of the Barak and the state of Assam.

Findings of the Study

1. Lack of awareness among the people about potential of the local resources for development.
2. Mismanagement in developmental works under the government departments.
3. Planning is imperfect and implementation incomplete in most cases.
4. Implementation is partial due to bureaucratic handlings.
5. Ground water resources in the GP area are not exploited to their full potential.
6. Area can be developed socio-economically by using the local resources.
7. Multiple crops can be grown in the area, provided water supply is ensured.
8. Area is also suitable for fishery and hatchery development.

9. Stagnation in economic development is due to poor planning and implementation.

10. Area is endowed with rich water resources for sustainable development.

11. Development of Pakai Beel will amount to development of the larger Narsingpur area.

Conclusion

Water is essential for life and the water bodies are useful to all living beings. This precious aquatic liquid should be harvested and preserved systematically to meet the contemporary challenges of environmental heterogeneity and global warming. Whereas water is so valuable, in the Barak Valley flash foods and water-logging in the monsoons have proved over the years to be the cause of misery and sufferings for the human beings and damages to habitats, plants and animals. On the other hand, the average rainfall in Northeast India has been recording a declining trend in the recent years. The scarcity of water for various usages of livelihood and for the economic activities like agriculture, horticulture, fishing etc. is an important concern in the region. These facts call for a proper system of watershed management to ensure the availability of water throughout the year. The people also have to learn to harvest and preserve water. The water resources of the area should be properly developed and utilized to their full potentials not only to resolve the problem of scarcity of water and to minimize the recurrence of flood and water-logging but also to ensure environmental preservations and sustainable economic growth.

REFERENCES

Sivakami, A. (2002), 'Uses of Water Resources – Ethics & Environment: A Case Study of Developing Countries', in: Kumar, A. (ed.), *Ecology & Conservation of Lakes, Reserviors & Rivers,* pp-210-223, ABD Publishers, Jaipur.

Kaushik, S. & Saksena, D.N. (1999). 'Physio-Chemical Limnology of Certain Waterbodies of Central India', in: Vijoy Kumar, K. (ed.) *Freshwater Eco System of India,* pp-1-7, Daya Publishing House, Delhi.

Gupta, A. (2007), 'Water Harvesting as a Management Strategy in Agriculture and Public Health: The Barak Valley Scenario', in: Roy, N. Biswas, D. and Dutta, B.K. (eds.), *Water Harvesting and Water utilization in North East India*, pp. 108-118, Published by Assam University, Silchar.

Sanghvi, Sheela, Pahwa Prem, S. (2001), 'Water Harvesting, Purification and Distribution Management', *The Degradation of Freshwater Supplies*, pp.13-32, Dominant Publishers and Distributors, New Delhi.

Shaxson, T. F. (2000), People's 'Involvment in Watershed Management: Lessons from working among Resource-Poor Farmers' in: Lal, R. (ed.), *Integrated Watershed Management in the Global Ecosystem*, pp. 345-361, CRC Press, LLC Florida, USA.

Abbasi, S.A. (2001), 'Water Resources Projects and Their Environmental Impact', *Approach to The Present Survey, Status of Environmental Impact, Assessment of Water Resources Projects in Krishna River Basin*, pp.173-203, Discovery Publishing House, New Delhi.

Rajora, R. (1998), 'Integrated Watershed Management', *Philosophy and Concept of Watershed*, pp. 27-43, Rawat Publications , Jaipur/New Delhi.

Statistical Handbook (2005), Govt. of Assam, Directorate of Economics and Statistics, Assam, Guwahati.

Guha, U.C. (1910), *Kacharer Itivritta*, 1st edn. 1910, reprint Sadana Library , Dhaca.

Sharma, S.L. (2006). 'Rethinking Sustainable Development – Ecology and Culture in India's Developmental Context, in: Choudhury, S.K. (ed.), *Culture, Ecology and Sustainable Development*, pp-27-43, Mittal Publications, New Delhi.

Khan, T.I (2001), 'Environmental Policies for Sustainable Development', *Environmental Policy Instruments*, pp.30-58, Pointer Publishers, Jaipur.

Salodia, P.K. (1996), 'Freshwater Biology An Ecological Approach, *Freshwater Ecology*, pp. 46-52, Surabhi Publications, Jaipur.

Borgstrom, R. (1994), 'Freshwater Ecology and Fisheries' in: Balakrishnan, M., Borgstrom, R. & Bie Stein, W. (eds.), *Tropical Ecosystems: A Synthesis of Tropical Ecology and Conservation*, pp. 41-69. Oxford & IBH Publishing Co. Pvt. Ltd., New Delhi.

11

Local People's Attitude towards the Barak River, Assam

Anuradha Singha & Ananta Pegu

[**Abstract:** River Barak of Assam - the second largest river in the north-eastern region of India - sustains millions of people. It is the perennial source of water for the people of Barak Valley providing ecological and economic services to the people. In spite of its immense importance, the Barak River is being increasingly polluted both by domestic sewage as well as industrial effluents. Annual flooding by the river has been aggravated by human activities such as deforestation leading to soil erosion, sand casting, sedimentation, etc. Global warming and climate change are also likely to alter the hydrological realm of the river. Water is a natural capital which needs to be conserved for the sustainable development of the valley. It should be understood that the government's action only will not be sufficient to mitigate the challenges unless it is accompanied by active co-operation of the local people. For this, it is necessary to analyze the level of people's concern about the conservation of the water resources. With this view point, the paper attempts to assess the attitude of the people of the Badarpur locality of Barak Valley towards the Barak River.]

Introduction

In today's world of industrialization with increasing global warming, the problem of sustainable development of water resources has emerged as an important issue to be addressed urgently. Water resources are highly sensitive to weather variation and climate change. Besides the natural forces, anthropogenic (human-made) factors also play an important role in water related issues in supplementing the effects of climate change. Half the watershed area of the world's fresh water systems is estimated to have been lost in the twentieth century (Murthy : 2004). From time immemorial, rivers have been valued highly by the people as these are the life sustaining gifts of nature. But, it is seen that with population growth and process of industrialization this scarce economic treasure is being depleted and polluted and also inappropriately used to a large extent over the last few decades. Side by side the increasing human activities, the climate change and global warming have aggravated the problem further, resulting in the accumulation of green house gasses like carbon-di-oxide, methane etc in the atmosphere affecting the pattern and nature (both quality and volume) of the water resources/rivers and thereby altering the frequencies of floods and droughts and causing various diseases all over the world.

Water resources are among the precious gifts of nature to the humankind. It is a vital necessity for the existence of life form on earth. But, in today's world of industrialization the people are causing systematic damages to these resources either directly or indirectly. The human induced global warming forms the most important factor of the climate change, bringing alterations in the frequencies of flood and droughts due to the changes in trend and pattern of the rainfall. During the summer the river gets over-flooded, while during winter it dries up and the drainage capacities are reduced due to silt deposition. Further, the problem of water contamination due to human activities is increasing significantly. People are indeed mainly responsible for water resource degradation. Hence, the water resource management for sustainable development can be meaningful only when the people are concerned and aware of their doings and the effects of such doings.

Assam, one of the largest north-eastern states of India, has also been witnessing environmental change which is basically engendered by anthropogenic factors. The state is topographically divided into the Brahmaputra valley floodplain and the Barak valley floodplain. The Brahmapura and the Barak are among the major rivers in the country. Assam is thus fortunate to be very rich in water resources. Indeed, water resource is considered as one of the most important resources of the state. But, due to the increasing human activities in the hills as well as in the plains of the region the water pollution, heavy floods, soil erosion, sedimentation or silt formation, etc have been accelerated in the state. The Barak Valley, which constitutes the southern division of Assam, comprises of three districts, namely, Cachar, Karimganj and Hailakandi. The valley is endowed with enormous water resources with various water bodies like lakes, swamps, ponds, etc, besides the large perennial river – Barak and her tributaries. The name Barak Valley is also derived from the river Barak.

The Barak River

Barak is the second largest river in Northeast India. The Barak basin is spread over India, Bangladesh and Burma and drains an area of 41,723 sq km in India which is nearly 1.38% of the total geographical area of the country. The Barak River has a total length of 902 kms, out of which the Indian reach is 564 kms. It rises in the mountains near the border of Manipur and Nagaland and forms a part of the northern boundary of the Manipur state with Nagaland where it is known as Kirong (State of Environment report, Assam, 2004). The river enters the plains of Barak Valley near Lakhipur and traverses the valley in a westerly direction up to a place called Bhanga near Badarpur, where it bifurcates into two branches, known as the Surma and the Kushiyara, which reunite near Bhairab Bazar in Bangladesh with the name of Meghna. The latter joins the Brahmaputra (locally known as the Padma in Bangladesh) and eventually flows into the Bay of Bengal. (Goswami : 2009). In India the basin spreads over the states of Meghalaya, Manipur, Mizoram, Assam, Tripura and Nagaland with the drainage area of 10650, 9567, 8866, 7224, 4688 and 728 sq. kms respectively (Central Water Commission). In India principal tributaries of Barak are Jiri, Chiri, Madhura, Jatinga, Katakhal, Dhaleshwari, Singla, Longai and Sonai.

Role of Barak in Life of the Valley

From time immemorial the mighty river Barak is feeding the people of Barak Valley. Although it is true that during the monsoon season it causes devastations to the masses in the form of flood, river bank erosion, sand casting, silt formation, etc. damaging agriculture, environment, human life and property, this river is the life-giving boon of nature to people of the valley. It plays a very important role in life of the people in terms of generating income and livelihood and in meeting the various household requirements.

It is the most important source of surface water in the valley. The assessed average annual surface water potential of the Barak is 585.6 cubic km. Out of this, only 24.0 cubic km is annually utilizable (Mahanta : 2006). The Barak is also significant for its household water supply throughout the year. This valley is predominantly an agricultural region, with 70% of the work force being engaged in agriculture and other primary sector activities. Water is the prime requisite for such activities and production. As the arrival of monsoon is irregular, there is immense importance of assured water supply in agricultural fields through irrigation system. The Barak is the only source of the assured water supply to the agricultural fields, basically, through the lift irrigation system where pump sets are utilized to draw water from the river.

The Barak basin has fertile soil where various seasonal crops, vegetables, edible oils, etc. are cultivated basically by the people dwelling in the banks of the river. The cultivable area of the basin is about 13.04 mha, which is 7.1% of the total cultivable area of the country (Mahanta : 2006). Apart from serving the personal needs, such kind of cultivation helps in income generation when the stuff is brought to the market. Fishing is another activity practiced in Barak. Across the entire river the commercial fishing is practiced and a good amount of income is generated from the fisheries. Again, some rural people in the valley, who belong to landless and economically backward section, are meeting their requirements of fish by indigenous methods of fishing from the Barak and sometimes selling them out for supplementing their income for livelihood. There is a huge potential of fishery development in the valley, although it is not utilized properly. Navigation or transportation is another service provided by the river Barak. The river Barak and its tributaries are extensively used by the local communities for localized transport system. It provides the readymade route and the cheapest mode of transportation of various cargos and commodities. There are vast opportunities for developing inland water transport in the valley in these days of high transportation cost and crisis of oil. The Barak has also a great hydro-power potential. As per the latest assessment, the hydro-power potential in the basin is 2042 MW at 60% load factor. However, the bulk potential remains yet to be explored. (Central Water Commission)

Apart from the above mentioned economic importance of the Barak, the river also has ecological significance. The Barak basin has an ample forest cover which helps in maintaining the ecological balances by preventing soil erosion. Further, Barak has great potential of adventure sports and of water tourism and recreation opportunities. It is indeed sad that these have not been harnessed properly.

Problems related to River Barak

The major problems which the River Barak has been facing are flood, river bank erosion, sand casting, and silt deposition. Settlements, infrastructure, thousands of hectares of fertile land, crops, etc. have been lost to the river bank erosion and due to frequent shifting of the course of the river. This problem is aggravated by deforestation in the riparian zone for establishing new human settlements and for various developmental purposes. Though flood, erosion, etc. are annual phenomenon of the river, the extent of occurrence of flood and erosion has been intensified during the last few decades bringing misery to the people of the valley; loss of life and property and adverse effects on the economy of the region as a whole. These effects are further aggravated by the global warming and climatic change, which are responsible for altering the hydrological realms of the river.

Apart from the natural forces of climate change, the anthropogenic factors play a very important role in inviting such calamites. It is the human who in the name of development and economic growth is ignoring the very basis of his life, i.e., the environment. Human is violating nature's rule and exceeding the limit of earth's own ability to absorb and purify and to maintain the ecological balance, thereby creating a dangerous and harmful situation of environmental imbalance in the region. The activities of the human have led to the accumulation of green house gases like carbon-di-oxide, methane, carbon monoxide, etc. in the atmosphere and the earth is becoming warmer, affecting the pattern and nature of the climate, which in turn is altering the frequencies of floods and droughts.

The river gives fresh water to the people and takes within it all the waste products. However, the culture of dumping all kinds of waste materials and effluents is increasing day by day, due to the increase in population, exceeding the carrying capacity of the river. Various effluents and sewages from households and industries are released into Barak, contaminating the river water and making it unsuitable for various household purposes, besides leading to the spread of various water borne diseases. The degradation in water quality is also affecting the survival of aquatic life in the river. It is affecting the fishing activity and its commercialization. Further, the establishment of industries near the river has been degrading the water quality to a large extent due to the dumping of sewage, waste products and chemicals from the industry. As for example, the HPC paper mill in Panchgram and the BVCL industry are known to be continuously affecting the water quality of the Barak. Sometimes, immersion of idols in the Barak also causes damage to the water quality. In Barak Valley, after the Pujas, the idols of gods and goddesses are freely immersed in the rivers. This happens in a large-scale after the Durga Puja (autumn festival] which is the most important festival in the valley. Various chemicals and heavy metals which are used in making the idols as well as many plastic and other insoluble ornaments that are put on the idols for decorative purposes invariably produce contaminating effects in the river, resulting in water pollution and damages to aquatic life.

Objectives of the Study

Human is mainly responsible for environmental degradation, climate change and various water resource related problems. The phenomenal population growth in the valley, along with the desire of the people to attain higher economic growth and development, is imposing heavy toll on the water bodies. The River Barak, which should be conserved as the main source for meeting the water demands of the people of Barak Valley throughout the year, is indeed worst affected by the devastating effects of population explosion in the valley. In this regard, the government alone can not bring the desirable changes unless it is supported by the local people of the valley. Each and every person in the valley have to be aware of their responsibility to protect the river as the premier water resource for sustainable development of the region. Sustainable development means the development that meets the needs of the present generation without compromising the ability of the future generations to meet their own needs (Rao : 2000). As it is the people who can protect or destroy the sources of their own livelihood, it is very important to assess the people's attitude towards these resources. In other words, those who are the users and beneficiaries of the river and residing near its banks. Hence, the study is intended to make an assessment of the attitude of the local people of Barak Valley, particularly those who live near the banks of the river, towards the River Barak and their awareness about is conservation.

Methodology

For the purpose of the study, information has been collected from sixty-four households residing in the immediate bank of Barak in the Dighirpar village in Badarpur area under the Karimganj district. For gathering information, the face to face interviews were conducted in the village. A simple tabulation and percentage calculation have been done in analyzing the data.

Findings

In order to get the people's opinion, only four questions were presented to the households and they were asked to reply in the form of 'agree', 'disagree', or 'neutral'. These are:

(i). Question 1: for the economic growth and employment generation, should government set up any industry near the Barak river, which may pollute the river water?

(ii). Question 2: should the people pollute the water of Barak for their vested interests?

(iii). Question 3: should the present generation take any step, by sacrificing their time and money, for conservation of river water for the future generation?

(iv). Question 4: is it necessary to save the River Dolphin?

Table 1: Household attitude towards the River Barak

Questions	*Number of households*		
	Agree	*Disagree*	*Neutral*
Question 1	52 (81.25%)	10 (15.62%)	2 (3.12%)
Question 2	20 (31.25%)	44 (68.75%)	——
Question 3	58 (90.62%)	4 (6.25%)	2 (3.12%)
Question 4	56 (87.50%)	7 (10.93%)	1 (1.56%)

As presented in Table 1, out of the total, 81.25% households agreed that the setting up of industries is essential for employment and income generation even if the river gets little polluted, while only 15.62% households disagreed to the setting up of industries near the river and 3.12% were neutral. This shows that most of the households were ready to accept the water pollution in return of the employment and income generation for the people, i.e. for economic growth and development. This means that for the people in the area the problem of unemployment and lack of income is more serious a problem than the water contamination.

About the second question, only 31.25% opined that the people may pollute the river for their vested interests. On the other hand, 68.75% disagreed to such dispensation. This particularly gives the impression that most of the households are pro-environment and they are aware of the limit to which they can use the river. Most of them are aware of the importance of the river to their survival.

When the third question was asked, 90.62% of the households agreed that the present generation should take appropriate steps for the river water management, sacrificing their time and money for the sake of the future generations. Whereas, only 6.25% disagreed to sacrificing interests of the present generation for the sake of future generations and 3.12% of households showed a neutral attitude in this regard. This shows that majority of the households are concerned about the degrading water quality and the problem of pollution and contamination of the river. They think that as they are getting benefits from the river, the future generation should also have the opportunity to enjoy those benefits. This reflects the awareness of the local people about the sustainable development of the river resource.

To the fourth question, i.e. the Barak Dolphin, 87.50% households think that it is necessary to protect the Dolphins. Only 10.93% think that it is useless to protect the Dolphins and 1.56% was neutral or unconcerned about the issue. This implies that majority of the households have strong desire to protect the Dolphins.

In addition to the above, the households were also asked to state the purposes for which they want the River Barak to exist. The responses are presented in Table 2.

Table 2: Ranking of the Utilities of the River Barak by the Households

Aspects of utility of the Barak River	*Number of Households*	*Percentage*	*Importance rank*
1. Using the water for households/irrigation	22	34.37%	1
2. Fishing	10	15.62%	4
3. Using the River as sink (dumping waste)	2	3.12%	5
4. River transport	19	29.68%	2
5. Natural beauty/ ecological services.	11	17.18%	3

From Table 2, it may be clear that the most valued service of the River Barak to the people of the valley is its role in the supply of water for the household consumptions and irrigation; 34.37% households valued it. Second is the river transportation, with 29.68% respondents voting for it. Natural beauty/ecological services and fishing came third (17.18%) and fourth (15.62%) respectively. Lastly, only two households (3.12%) voted for using the river as a sink. Thus, it is clear that the most desired service of the river is the use of the river water for household and irrigation purposes.

Conclusion

The people of the Dighirpar village in Badarpur area in the Karimganj district are aware of the effects of any kind of undesirable activity towards the River Barak. Here, both the pro-environment and the environment-neutral opinions were recorded. The opinions were mostly pro-environment. It seems that the people are aware about the importance of the Barak and the necessity of its appropriate management in the interest of the future generations of the valley. The signs of awareness and concern among the people for the protection of the river from harmful activities were clearly discernable. This indicates that the general attitude is in favour of sustainable development of the Barak River. However, some contradiction arises when the case of the economic growth and development is presented to the people. Being economically backward and because of the lack of opportunity for employment and income, the people in the Barak Valley desire that the economic growth and development should take place at a faster rate. So, sometimes they prefer economic growth and development by establishing mills and industries at the cost of environmental or water resource degradation.

As both economic growth and sustainable water management are important, the cost-benefit analysis of the setting-up of industries and the values of the environmental resources may show the way to an acceptable solution. The option for environment friendly alternative schemes of development should be the approach for the economic growth and poverty resolution in the region. The spread of environmental education among the local people, along with the awareness campaign regarding the problems of conservation of the water resources of the valley, would be helpful in sensitizing the people and achieving sustainable development of the water resources in Barak Valley.

REFERENCES

Brahmaputra and Barak basin. Central Water Commission (http://www.cwc.nic.in)

Goswami, Dulal C. (2009), *Management of Water Resources of North-East India: Need for an Integrated Regional Plan* (www.indianfolklore.org)

Mahanta, Chandan (2006), *Water Resources of the North East: State of the Knowledge Base*, Background paper No. 2, Indian Institute of Technology, Guwahati, India.

Murthy, D.B.N. (2004), *Environmental Awareness and Protection*, Deep and Deep Publications Pvt. Ltd., New Delhi.

Rao, P.K. (2000), *Sustainable Development*, Blackwell Publishers, London.

State of Environment Report Assam-2004, Assam State Technology and Environment Council.

12

Big Dams – Bigger Threats

Parthankar Choudhury, Pijush Kanti Das & Projit Kumar Palit

[**Abstract:** Barak Valley at times gets flooded with water; while during other times suffer for water scarcity, both for drinking and irrigation. Barak is the main river which maintains the balance of peripheral water bodies including lentic ecosystems. Therefore, all plans for sustainable water usage should be initiated considering these facts in the backdrop. This paper deals with few such pertinent aspects as a big dam is likely to come up in its upstream, which may jeopardise the entire biota and environment in downstream areas. It lays emphasis on an extensive downstream environmental impact study from the proposed dam site up to sea-mouth with experts from Non-Government Organisations, particularly from environmental outfits, IITs and Universities in the affected countries, to assess the possible detrimental impact on the environment and life of inhabitants in the catchments areas at large.

During recent days, lot of hue and cry have been raised against the construction of dams at Tipaimukh and Subansiri. The protests have been registered from Manipur and the Barak and Brahmaputra valleys of Assam, besides those from Bangladesh. All these protests should be viewed from pro-environment and human point of view. Both Subansiri and Tipaimukh dams are at the confluence of Indo-Myanmar, Indo-Malayan and Indo-Chinese Biodiversity hotspot zone. These areas are characterised by the presence of a large variety of plant and animal species, many of which are not seen in rest of the world. A large number of them have been categorised as endangered and threatened as per the IUCN Red Data book and the Wildlife (Protection) Act, 1972. Once the dams are constructed, these innocent and endangered species would have no other alternative but to perish! Under such a situation, does the construction of dams in the proposed sites speak well in favour of Biodiversity conservation? It is a matter of deep concern that the rock filled 500 metre long and 162.8 metre high Tipaimukh dam is to be constructed at the earthquake zone-V, wherein there will be constant pressure of water. If for any reason it cracks, the entire civilization of the whole of downstream will be washed down in no time. The age old Barak-Surma culture will live in history only. No force or technology can ensure against such catastrophic mishap. Besides the above mentioned issues, other impacts like water scarcity, crop cultivation, navigation, siltation, ecological imbalance, river pollution, extinction of aquatic life forms and the like are never-the-less important frontier areas that deserve careful and serious attention before construction of the dam. Keeping all these facts in view, the present paper aims to analyse the entire issue from the pro-environment and humanistic point of view and advocates against the construction of the dams, the life time curse for the inhabitants of Barak-Brahmaputra-Surma basin. The key words used are Biodiversity, Downstream areas, Environmental impact study, Subansiri Dam, and Tipaimukh Dam.]

Introduction

River means a wide natural stream of water that flows in a free flow channel with more or less well-defined banks. The prosperity of India to a large extent depends on her river systems. They flow down in various directions seeking the level, cutting valleys, sometimes through mountain ranges, sometimes on land and occasionally changing their courses. The *Markandeya Purana* (LVH.30) says, *"All the rivers are sacred, all flow towards sea. All are like mothers to the world, all purge away sins"*. During later Vedic period, the Aryans occupied the whole of India, and they conceptualized the rivers by seven principal streams, called the Ganga, Yamuna, Godavari, Saraswati, Narmada, Sindhu and Kaveri. The hymn thus goes, *"Gange cho Yamune chaibo Godavari Saraswati : Narmada Sindhu Kaveri Jale'smin Samadhim kuru"*. The glimpses of the ancient history and culture of India thus speak about beneficial properties of rivers and the need for their conservation (Palit : 2006). Nowhere in the ancient records of India, has any attempt to disrupt the free flow of rivers been noticed.

The proposed Tipaimukh Dam is to be located 500 metres downstream from the confluence of Barak (or Tuilong) and Tuivai rivers, and lies on the south-western corner of Manipur State (24°142 N and 93°132 E approximately). It is a huge earth dam (rock-fill with central impervious core) having an altitude of about 180 M above the sea-level with a maximum reservoir level of 178m and 136m as the MDDL (minimum draw down level). The dam was originally conceived to contain only the flood water in the Cachar plains of Assam. Later on, the emphasis has been placed on hydroelectric power generation, having an installation capacity of 1500MW with only a firm generation of 412MW (less than 30 per cent of installed capacity) (Anon. : 2002). In order to appease the people of Manipur state, the project proponent – NEEPCO - has been building up a list of benefits from the dam that include high-class tourism, free power sharing, resettlement and rehabilitation package and an all round rosy picture of development.

Over the past few decades, the issue of Tipaimukh Dam has created a lot of disenchantment in regard to scientific, technical, economic and environmental feasibility of the dam, especially concerning the states of Manipur and Assam and neighbouring Bangladesh. In the light of disillusionment over the past few years as a consequence of ongoing construction of Subansiri Dam in the Brahmaputra valley, this paper makes an attempt to provide a brief account of the possible impact of the proposed Tipaimukh Dam on the ambient environment and tries to analyse the entire issue from the pro-environment and pro-human point of view. Such a consideration would reveal the nature and extent of the variety of risks being taken by constructing a mega-dam at Tipaimukh.

Tipaimukh Dam

The proposed Tipaimukh Dam would be constructed 500 meters downstream from the confluence of the Barak and Tuivai rivers in the southwestern corner of Manipur. The site of dam construction is located at 24°142 N Latitude and 93°1.32 E Longitude approximately and the altitude is 180 m above MSL.

(i) **Particulars of Tipaimukh Dam**: The dam would be an earthen-rock filled dam. The proposed length is 390 meters, estimated height is 162.8 meters and its maximum reservoir level is 178 meters. Total area required for construction of the dam, including submergence area, is 30860 hectares. The implementing agencies are NHPC, SUTLAZ and the Manipur Government. The project cost, as estimated by the NEEPCO, was Rs. 5225.70 crores. However, the present revised cost of the project is Rs 10,000 crores (Anon : 2002).

(ii) **Purpose of the Dam**: The dam was originally designed to control flood-waters in the Barak Valley. However, Hydropower generation was later on incorporated into the project. The project would have an installation capacity of 6 × 250 = 1500 MW but what is more important is that it would have a firm generation of 412 MW only.

Power Generation - A Boon or Curse in Disguise?

It may be recalled that the idea of a dam for the river Barak, originally called Barak Dam, is very old. It was first conceptualized in 1954 to contain floods. The decision to locate it in Tipaimukh was also guided by its original idea of controlling floods. The project was put in the cold storage for a long time because it did not have the requisite environmental and management plans. Finally, the Brahmaputra Board jumped into the scene. The Board carried out studies and revised the plan as a multi-purpose project. The same has been re-revised by NEEPCO. In the meantime, the fund involvement for the project has gone up many times. On the other hand, the installed capacity of 1,500 MW is supposedly for the monsoons when the river is in high spate, whereas at any given time it is usually only 412 MW. The power generation is undoubtedly important in modern conditions, but the decision for setting up a mega dam, like the Tipaimukh Dam, should be taken only after due consideration of all implications, including lessons of history.

Lessons from Subansiri

(a) **Influence of Subansiri on livelihood sectors**: Before the construction of Subansiri Dam, the situation that prevailed there was quite good and people from various walks of life earned their livelihood under the then conditions. The tangible benefits obtained by the local communities are given in Table 1.

Table 1

	Influence of Subansiri river on livelihood sectors of the downstream people	*Benefits*
1	Agriculture in the downstream of Subansiri	Food/Income
2	Collection of wood in the downstream of Subansiri	Income
3	Livestock (buffalo & battle) rearing in the downstream of Subansiri	Income
4	Fisheries in the downstream of Subansiri	Food/Income
5	Transportation in the downstream of Subansiri	Income
6	Trade in the downstream of Subansiri	Income
7	Riparian economic flora in the downstream of Subansiri income	Economic values/
8	Sand and Gravel mining in the downstream of Subansiri	Income

The site where the Subansiri Dam is being constructed was the abode of a variety of flora and fauna. Being located in the biodiversity rich zone, a variety of wild flowers and several rare orchids occurred in the area. In the adjoining Dulong Reserve Forest quite a good number of endangered and rare species of butterfly used to flourish. Fish diversity in the Subansiri drainage system include different species of *Mystus, Tor tor, Tor putitora, Labeo rohita, Labeo calbasu, Puntius sarana sarana, Labeo bata, Aspidoparia jaya, Cirrhinus reba, etc.* Besides, a lot of hill stream cold water fish diversity is common in the Subansiri drainage system. *Labeo dero, Labeo pungusia etc* and the Gangetic river dolphin (*Platinista gangetica gangetica*) are among them. The species that prefer to live in deep gorge also flourished in the Subansiri basin. The construction of a dam in such a beautiful paradise of plants and animals may bring about drastic changes in the habitat of all these species.

(b) **Biodiversity Scenario**

(i) **Flora:** Studies carried out on the vegetation of the area reveals that the area is generally varied and botanically interesting. It has been observed that the vegetation cover in the study area varies from tropical to sub-tropical evergreen moist deciduous forests, bamboo forests and grasslands and riparian forests. The predominant plant species that are found in the area include *Ficus sp., Macaranga sp., Duabanga sp., Lagerstroemias sp., Bombax sp.,*etc. Various fern species that occur are *Cyathea gigantean, Angiopteris evecta, Pteris sp.* etc. Timber species include, *Dipterocarpus sp., Terminalius sp., Michelias sp., Artocarpus sp., Kayaassamica sp., Tetrameles sp., Duabanga sp., Gynocordia odorata, Aquilaria agallocha* and *Tree fern* etc. The epiphytic flora is very rich and diverse and include *Dischidia sp., Hoya sp., Pothos sp., Raphidophora sp.,* etc. Orchids of the species *Dendrobium* and *Cymbidium* are quite abundant. Thus the area represents the gene bank of a good number of plant species (Choudhury and Das : 2010). Of these, the species that deserve special attention are *Michelias sp., Magnolias sp., Bucklandia populanea, Podocarpus nerifolia, Cyathea gigantean, Cycads, Caryota sp., Licula sp., Pinaga sp.,*and rare ochids like *Lilium macklieae, Paphiopedilum spicerianum,* and *Paphiopodium villosum.* Once the dam is constructed, the stagnant water may favour the growth of aggressive weeds like Water hyacinth, *Eichhornia crassipes, Zizania sp.* etc. which will be very destructive and may create 'Phumdi' like structures in the reservoir water as in Loktak lake of Manipur (Choudhury & Das : 2010). This aspect also deserves careful attention.

(ii) **Fauna:** The catchments area is very rich in the variety of wild birds, reptiles and mammals. The important animal species include Clouded Leopard, Malaya Sun bear (Honey bear), Golden cat, Marbled cat, Stump tailed monkey, Slow loris, Goral, Hoolock gibbon, Capped langur, Python, Blyth's tragopan, Peacock pheasant, Indian pied hornbill, Rufus necked hornbill, Brown backed hornbill, Chinese pangolin, Indian wild dog, Leopard etc among many others. Important aquatic fauna include *Gangetic river dolphin, Hilsa,* and various sp. of Mahaseer etc. Construction of a big dam in the site may lead to their habitat loss. On the other hand, translocation of the species living therein to some other area is again a subject to be dealt with seriously, and the feasibility need to be explored.

Earthquake Scenario

Northeast India figures as one of the six major seismically-active zones in the world. The other zones are California, Japan, Mexico, Taiwan and Turkey.

(i) **Tectonic Setting of Northeast India**: Northeast India and its adjoining areas have the following tectonic se- up and experienced major earthquakes as given in Table 2 (A & B).

Table 2(A) : Five Seismotectonic Zones

Sl No	*Area*	*Year (Richter Scale)*
1	Eastern Himalayan Collision Zone Two Earthquakes	1941 (M7.1), 1947 (M7.8)
2	Indo-Myanmar Subduction Zone	10 Earthquakes > 7.0 (during last 100 years)
3	Syntaxis Zone Great Assam Earthquake	1950 (M8.7)
4	Shillong Plateau, Mikir Hills and Assam Valley Zone Shillong Earthquake	1897 (M8.7)
5	Bengal Basin and Tripura-Mizoram Fold Belt Srimongal Earthquake	1918 (M7.6)
6	Cachar Earthquake	1984 (M5.6)

Table 2(B): Earthquake Incidence

Sl No	*Place*	*Year*	*Magnitude (Richter Scale)*
1	Cachar	21-Mar-1869	7.8
2	Shillong Plateau	12-Jun-1897	8.7
3	Sibshagar	31-Aug-1906	7.0
4	Myanmar	12-Dec-1908	7.5
5	Srimangal	08-Jul-1918	7.6
6	SW Asam	09-Sep-1923	7.1
7	Dhubri	02-Jul-1930	7.1
8	Assam	27-Jan-1931	7.6
9	Nagaland	1932	7.0
10	NE Assam	23-Oct-1943	7.2
11	Arunachal	07-Jul-1947	7.5
12	Upper Assam	29-Jul-1949	7.6
13	Upper Assam	15-Aug-1950	8.7
14	Arunachal	1950	7.0
15	Manipur-Burma	1954	7.4
16	Darjeeling	1959	7.5
17	Indo-Cachar	11-Nov-1984	5.8
18	Indo-Myanmar	06-Aug-1988	7.5

Impacts of Proposed Tipaimukh Dam

The total area required for construction, including submergence area, is 30860 hectares. Out of this, 20,797 hectares is forest land, 1,195 hectares is village land, 6,160 hectares is horticultural land, and 2,525 hectares is agricultural land. As per estimates of the authorities, the project will totally affect 311 sq. km. and 8 villages with 1461 families altogether.

(a) Downstream Impact

(i) **Impacts of the Dam in Tamenglong**: In Tamenglong district of Manipur, a total land area of 288.60 sq. km. is likely to be submerged. The Barak waterfalls and Zeliet lakes, which are linked with the history of the local people of Tamenglong, are likely to be submerged forever and the folklore and legends will have no monuments of proof and will live in history for the future generations. Once the dam is constructed, more than 40,000 people will be rendered landless. Sixteen villages situated on both sides of the river Barak will be permanently inundated. In addition to the eight villages as mentioned above, more than 90 villages, mostly of Tamenglong district, will be adversely affected. About 27.242 hectares of cultivable land will be lost. The township of Nungba (sub-division headquarters) and the villages along the NH-53 will be severely affected. The dam is likely to bring health hazard, water borne diseases, industrial pollution and many other environmental and ecological problems due to increase of water surface. Prolonged stagnant water will be unsafe for drinking, thereby may cause a lot of problems to the people (Sethi : 2006). Local people are of the opinion that the decision for construction of a dam in such an area was taken without a proper downstream ecological study which, they foresee, would create a big trouble in the future.

(ii) **Impacts on others** : The Zeilangrong Naga area stands to be submerged by the dam, besides being affected culturally. In a situation where the environmental impact assessment has not yet been carried out, the people have to depend on NEEPCO's claim that only four Naga villages in Tamenglong will be submerged. The ground reality is that altogether ninety villages will be affected, some more and some to a lesser extent.

(b) **Environmental impacts**: The environmental impacts of dams can be generally classified into two categories: (i) Those due to existence of the dam and reservoir; and (ii) Those due to the pattern of dam operation.

(i) **Due to existence of dam and reservoir** : Imposition of a reservoir in a river valley would obviously cause loss of habitats. Changes in downstream morphology of riverbed, delta, and coastline due to altered sediment load would result in increased soil erosion. Changes in downstream water quality cannot be ruled out as there would be changes in river-water temperature, increase of nutrient load, turbidity, dissolved gases, concentration of heavy metals, minerals and other contaminants. Reduction of biodiversity is another threat as this may happen due to blocking of movement of organism like Turtle, Hilsa, River Dolphin and other important endemic aquatic species living abundantly in the aquatic set up of river Barak.

(ii) **Due to pattern of dam operation** : The impacts that may take place due to the dam pattern include changes in downstream hydrology, changes in total flows, and changes in seasonal flows (e.g. flood may occur in spring or in winter). Short-term fluctuation in flows like change in extreme high and low flows, changes in peripheral water bodies, changes in downstream water (both quality and quantity), changes in riparian/floodplain habitats, livelihood and diversity pattern, total collapse of navigation, fishing and similar other activities and changes in tea, paddy and other economic crops.

(c) **Social and Economic Impacts**: There may be serious impact of the dam on economic, social and cultural life of the people and may call for relocation of communities. The local community is likely to lose control over water since there would be transfer of control from local level to the hands of dam controlling authorities. Besides, this would result in decrease in employment opportunity in agricultural sector and consequently average income of the farmers is expected to go down. In addition, there would be damage of homes and other institutions, and the homeless tribal people might have to change their age old occupations. Education for all, the aspect on which much emphasis is laid during recent decades, would take a volta-face scenario as there would be difficulties in schooling of the children. The other aspects, like change in land ownership and standard of living of the local tribal people, also cannot be ignored.

The total population of the Zeliangrong tribe in the Tamenglong district, according to the 2001 census, was 120,000.

The villagers spend 70-75 percent of their money in sending children to the city for education and, as a result of this, the literacy rate has gone up to over 65 per cent.

The most important ethical consideration that needs to be mentioned here is that the idea of Zeliangrong Naga as separate from others is based on the beliefs centered on Zeihlat and six other lakes. According to them, if the lakes go or the falls disappear, they would treat it like the people in the Gangetic valley losing Varanasi, Allahabad and Haridwar (Sethi : 2006).

Threats to Environment

Every hydro-electric project unavoidably submerges some of its ambient areas often containing valuable forests and its wildlife, cultivable areas, human habitations etc and thus affect the overall environment of the locality. It thus demands a careful scrutiny and assessment of the role played by the endangered forests in maintaining ecological and environmental balance and in meeting the needs of the indigenous communities for generations. The construction activities involving new settlement of the immigrant labour force, construction of roads, movement of heavy machineries, besides deforestation, may cause adverse environmental impact on a large scale.

The changes in physical and climatic conditions of the project site may also cause instability in the form of landslides, soil erosion and disturbance of pristine areas. Eroded soil might fill up the reservoirs to a large extent. It is anticipated that the variations in nutrient contents and dissolved oxygen might render the water inhospitable to aquatic life. The dam is likely to affect riparian communities by reduced capacity for self regeneration, reduced recharge of groundwater aquifers, enhanced pollution levels etc. Submergence due to constructions may be the step through which fragile land topography, many delicate plants, faunal population and tiny living organisms such as butterflies have to suffer great losses.

Discussion

Under the physiographic setup of Northeast India the hydroelectric projects have to be generally located in the hilly forest areas. Thus, the greatest impact is obviously on the forests. Therefore, Environment Impact Assessment (EIA) of the entire downstream area should be an essential and important pre-requisite for planning such projects. Due to the construction of the Tipaimukh Dam, a large part of the reserve forests in Manipur and some parts of reserve forests in Mizoram will be submerged. This deserves careful attention. Forests in the Northeast are generally very rich in plant and animal genetic resources. Many are endemic and some others are closely related to the species of adjoining countries. As the region still remains unexplored (both in terms of flora and fauna), special attention is needed to safeguard the endemic genetic resources.

In the light of the foregoing account, it is emphasized that construction of the Tipaimukh Dam should be taken up only after

Fig. 1 : Proposed Site for Construction of the Tipaimukh Dam

Fig. 2 : Subansiri Dam Site [Before Construction]

Fig. 3 : Subansiri Dam Site [After construction has started]

obtaining the free, prior and informed consent of the affected peoples both in upstream and downstream portions of the Barak River. The lessons learnt from big dams constructed elsewhere should open our eyes and impel us to think comprehensively on another identical dam at Tipaimukh. Or else, let us keep the river Barak in free flowing conditions for the generations to come.

REFERENCES

Anon. (2002), 'NEEPCO - 'A note on Tipaimukh'. Pp. 1-6. http://www.infraline.com/ power/setup/ Neepco/ TipaimukhHEP.

Choudhury, P. and Das, P. K. (2010), "Big Dams – Boon or curse in disguise – Tipaimukh and Subansiri experiences" in *Pragyan* (ISSN No. 0976-4283) – A Quarterly Journal of Academic, Intellectual and Career Pursuit, Published from Tinsukia College, Assam, Vol. VIII (No. 1), pp. 45-50.

Palit, P. K. (2007), 'The History of Rivers in ancient Northern India', in N. Roy, B. K. Dutta & D.Biswas (ed.), *Water Harvesting and Water Utilization in North-east India*, Assam University Publication, Silchar, pp. 76-83.

Sethi, N. (2006), 'Tipaimukh Dam in Manipur driving a wedge?', *Down to Earth* (Oct, 15, 2006), Centre for Science and Environment, New Delhi, Vol. 16 (10).

13

Barak Dam: An Overview

Bhaskar Singha

[**Abstract:** Energy plays an important role in the socio-economic life of the people and the development of a country or region. Water is one of the important sources of electric energy. With the economic transition taking place throughout the country, the demand for energy has increased many folds. The river Barak rises in India's Northeast and flows into Bangladesh to form the Meghna. The Tipaimukh hydropower project is planned as a dam at the junction of Assam, Mizoram and Manipur. The objectives of the proposed dam are to control the annual and devastating floods in the southern part of Assam and the power generation by NEEPCO. The present paper is devoted to explore the possible positive and normative aspects of the dam.]

Introduction

The demand for water and its related services has increased with growing population and rising economic status of the people. Dams are seen as effective instruments of meeting water and energy needs of the societies. The reason behind construction of large dams is to generate hydropower, besides providing continuous flow of water for irrigation and domestic purposes. However, in recent years, many questions are raised about the existing dams. It is, therefore, becoming difficult to plan new dams. These projects are often multipurpose so as to meet various needs such as flood control, irrigation and generation of hydropower. There are about 37,500 dams in the world of over 15m height. The population growth and improved living standards have created additional demand for food production, industrialization and economic growth. On the other hand, the people opposed to dams do commonly argue that dams cause severe damage to the environment. To maximize the positive impacts of the dams and to mitigate the negative environmental, social and economic consequences, the sustainability of water resource projects is required to be ensured.

Tipaimukh Dam

Energy plays an important role in socio-economic development of a country. Hydroelectricity in India is already well established. India ranks fifth in the world in terms of hydropower potential as per a report prepared by the National Hydroelectric Power Corporation of India (NHPC). Hydroelectric contributes 22% (21,104 MW out of 96,803 MW) of the country's total installed electricity generating capacity. 9% of the dams in the world are in India. In Manipur, Tipaimukh HE (Multipurpose) Project (1500 MW) has been planned at the junction of Mizoram, Manipur and Assam. An area of about 390 sq. km. is proposed for the construction of a rock-filled earthen dam at a height of 162 meters. The location of the dam is near about the Tipaimukh village in the Tipaimukh area of Churachandpur district (Manipur), about 500 meters downstream of the confluence of Tuivai River and Barak River. Thus, the name of the project was chosen to be Tipaimukh Hydro Electric Project. The dam project which was initially conceived to control flood ultimately became a power generation project as well, besides flood mitigation. According to official statements, this power project is to generate an installed capacity of 1500 MW. The height for the proposed dam shall be 162.8 meters. The estimated cost of the dam is about Rs 6,714 crores. The main objective of the proposed Tipaimukh dam is to control the fury of annual flood in the Barak Valley of Assam and to generate hydro power. The targeted capacity for generation of power is 1500 MW. The dam is supposed to be completed by the year 2012. The total area required for construction of the dam, including submergence area, is 30,860 hectares, of which 20,797 hectares is forest land, 1,195 hectares village land, 6,160 hectares horticultural land, and 2,525 hectares agricultural land. As per estimates of the authorities themselves, the project will totally affect 311 sq. km of land, 8 villages, 1461 families (belonging to the Hmar tribe).

Impacts of the Dam

The Tipaimukh Dam (popularly called Barak Dam) is likely to have the following negative impacts which, it is said, are to be kept in mind in the process of construction of the dam:

- More than, 40,000 people will be rendered landless.
- Eight villages situated in the Barak Valley will be completely submerged.
- More than 90 villages (mostly in the Tamenglong district of Manipur) will be adversely affected.
- About 27,242 hectares of cultivable land will be lost.
- The township of Nungba (sub-divisional headquarters) and the several villages along NH-53 will be severely affected.
- Dam will bring health hazard, water borne diseases, industrial pollution and other environmental and ecological problems due to increase of water surface.
- Increase in salinity will render groundwater unsafe for drinking.
- Possibility of frequent occurrence of destructive earthquake in the area.
- Decision was taken without proper ecological studies.
- Construction will directly affect the livelihood of a large number of people and cause economic and financial crises.
- Displacement of people and destruction of the habitats will pose a grave threat to the vibrant democratic system and people's right to live.

- Project will submerge the exotic flora and fauna and rich gene pools as Manipur falls under one of the genetic hotspot zones of the world with rare biodiversity resources.
- There will be problems of displacement, resettlement, rehabilitation, repatriation and development.
- Construction of the dam will be a violation of the democratic form of governance and functioning. Therefore, the indigenous voices are to be considered.
- Barak waterfalls and Zeilad Lake, which are connected with history of the Zeliangrong people, will be underwater for ever and the folklore and legends will have no monument of proof and it will become a make-up story for the next generation.

The positive impacts of the dam are mainly on the socio-economic benefit to the host communities, and aesthetic and recreational values in and around the dam site. Some of the positive impacts of the hydroelectric power projects, such as the Tipaimukh dam, are as follows:

- Hydropower is clean, renewable, low-cost alternative to other energy sources.
- Hydropower can be a true answer to the continuously rising demand for energy in the cities, towns and villages.
- Hydropower is emission-free and helps the nation meet its clean air goals.
- Dam will suffice the power need of the region.
- During the periods of drought the dam will provide irrigating water for crops.

Seismicity

Northeast India is one of the highest earthquake potential area in the world due to its tectonic setting, i.e. seduction as well as collision plate convergence. Analysis of earthquake epicentres and magnitudes of 5M and above within 100-200 km radius of Tipaimukh dam site reveals several earthquakes in last 100-200 years. It is found that within 100 km radius of Tipaimukh, 2 earthquakes of +7M magnitude have taken place in last 150 years, the last one having been occurred in 1957 at an aerial distance of about 75km from the dam site.

Another important aspect of seismic activity is that the shallow earthquakes are more disastrous than the deeper ones even if magnitude is relatively low, since destructive surface waves can be quickly and easily propagated from the epicenter. The majority of the earthquakes that had taken place on the western side of Manipur are shallow (50 km focal depth or less) which is due to the tectonic setting of the Indo-Myanmar Range. Under these circumstances, whether it will be a wise decision to construct a huge dam needs to be thoroughly investigated and discussed.

Conclusion

The environmental impacts of the Barak Dam are likely to be in various areas and measures. It is perhaps difficult to anticipate all the possible impacts beforehand and determine what may the positive or negative effects in terms of cost-benefit analysis. Undoubtedly, there are real and potential benefits to be obtained from the project. Industrial development would gain speed; irrigation channels and food production will be improved as a result of the increase in electricity generation. The flood menace shall also be substantially reduced. On the other hand, it is also abundantly clear that the people living downstream shall be adversely affected by the floods and displacements. Therefore, it is important that a thorough and scientific assessment of the possible impacts of the Tipaimukh Multipurpose Hydroelectric Project is undertaken before the final decision is taken on the construction of the dam. The opinions of the experts should be invited on technical aspects. More importantly, to ensure the participation of the affected people in Manipur and Assam, and further down in Bangladesh, an opinion polling should precede any decision on the dam and the project should not commence without the prior and informed consent of the affected people in upstream and downstream portions of the Barak river. The assessment should include impact of the project on the social, cultural, health, environment and economic aspects.

REFERENCES

The Sangai Express, 7 February 2003; 'CM's dam statement flayed'.

The Imphal Free Press, 28 January 2003; 'Dam Opposition Opposed'.

The Sangai Express, 11 January 2003; 'AMUCO joins anti-dam chorus'.

McCartney,M, 'Living with dams: managing the environmental impacts', www.iwaponline.com.

Sait, M; et al., 'Positive and Negative Impacts of Dams on the Environment', www.dsi.gov.tr.

'Tipaimukh: In the Quest for Survival', www. epao.com.

Select Bibliography

Abbasi, S.A., 'Water Resources Projects and Their Environmental Impact', Approach to The Present Survey, Status of Environmental Impact Assessment of Water Resources Projects in Krishna River Basin, pp.173-203, Discovery Publishing House, New Delhi, 2001.

Ao, M., J.R.B. Alfred, & A. Gupta, 'Studies on some lotic systems in the North-eastern Hill regions of India', *Limlogica* (Berlin) 15(1), 1984.

Ao, Temsula, 'Land Ethics and Eco-Management Among the Ao-Nagas : With Special Reference to Chanki Village', *Northeast Researches*, Journal of the Institute of Northeast India Studies, Vol. I, Kolkata, 2010.

Ananda, N., *Indian Freshwater Microalgae*, Bishen Singh Mahendra Pal Singh, Dehradun, India, 1998.

Anonymous, 'NEEPCO - 'A note on Tipaimukh'. http:// www.infraline.com/ power/setup/ Neepco/ TipaimukhHEP, 2002.

Anonymous. 'The Arunachal Pradesh Water Resources Regulatory Authority Act, 2006 (Act No. 15 of 2006)', *The Arunachal Pradesh Gazette*, No. 106, Vol. XIV, December 15, 2006, Naharlagun. Government of Arunachal Pradesh, Law and Judicial Department, Itanagar, 2006.

Anonymous, *Report on Irrigation Scheme*, The Department of Irrigation, Assam, Silchar Divisional Office, Nagatila, Silchar, 2009.

Anonymous, *Report on Fishery Development*, The Directorate of Fishery, Assam, Guwahati, 2009.

APHA, *Standard Methods for the Examination of Water and*

Wastewater, 21st edn., APHA, AWWA, WPCP, Washington DC, USA. 2005.

Bagra, K., K. Kadu, K.N. Sharma, B.A. Laskar, U.K. Sarkar, & D.N. Das, 'Lists of species, Ichthyological survey and review of the checklist of fish fauna of Arunachal Pradesh, India', *Check List* 5(2), 2009.

Baruah, U.K, A.K. Bhagowati, R.K. Talukdar & P.K. Saharia, 'Beel Fisheries of Assam: Community-based Co-management Imperative', *NAGA*, The ICLARM Quarterly, 23 (2), 2000.

Battish, S.K., *Freshwater zooplankton of India*, Oxford and IBH Publishing Co. Pvt. Ltd, New Delhi, 1992.

Bezbaruah, M.P., 'Regional and Seasonal Variations in the Trends in Rice Production in Assam from 1974-75 to 1994-95' (Summary), *Indian Journal of Agricultural Economics*, Vol.52, No.3, July-September, 1997.

Bhagawati, P., *Fishery in Assam*, NEDFI Data bank, 2009.

Bhatt, S., 'Scoping exercise for mainstreaming the ecosystem approach into integrated water resource management, India', IUCN Water and Nature Initiative: Final Report, 2004. [http://app.iucn.org/dbtw-wpd/edocs/Rep-2004-018.pdf.].

Bhattacharjee, J.B., *Cachar Under British Rule in North East India*, Radiant Publishers, New Delhi, 1977.

Bhattacharjee, J.B. (ed.), *Development Strategies for Barak Valley (Assam)*, Akansha Publishing House, New Delhi, 2009.

Bhattacharjee, J.B. (ed.), *State and Economy in Pre-Colonial Manipur*, Akansha Publishing House, New Delhi, 2010.

Bhattacharjee, Sarbani, "Environmental Features of Subbuñga (Subong) as recorded in the Tipperâh Copper-plate of Lokanâtha [*circa* 7th Century AD]", *Northeast Researches*, Journal of the Institute of Northeast India Studies, vol. I, Kolkata, 2010.

Biswas, A.K., 'Integrated Water Resources Management (IWRM): A Reassessment', *Water International* 29 (2), 2004.

Bhuiyan, J.R. & S. Gupta, 'A comparative hydrobiological study of a few ponds of Barak Valley, Assam, and their role as sustainable water resources', *J. Environ. Biol.*, Vol. 28, No. 4, 2007.

Bohra, P.H., 'Water Harvesting: a Suggestion', *Yojana*, July 2001.

Borgstrom, R., 'Freshwater Ecology and Fisheries' in: M. Balakrishnan, R. Borgstrom & W. Bie Stein (eds.), *Tropical Ecosystems: A Synthesis of Tropical Ecology and Conservation*, pp. 41-69. Oxford & IBH Publishing Co. Pvt. Ltd., New Delhi, 1994.

Brahmaputra Board, *Water vision. North East 2050. Report of the Brahmaputra Board*, Ministry of Water Resources, Govt. of India, New Delhi, 2000.

Brahmaputra and Barak basin. Central Water Commission (http://www.cwc.nic.in).

Carruthers, Ian, 'Irrigation under threat – a warning', *Asian regional Symposium*, Vol. 11, Hydraulics Research, Walling Ford, England, 1988.

Chakraborti, D., E.J.K. Singh, B. Das, B.A. Shah, M.A. Hossain, B. Nayak, S. Ahmed, & N.R. Singh, 'Groundwater arsenic contamination in Manipur, one of the seven North- Eastern Hill States of India: a future danger', *Environ Geol.* 56, 2009.

Chakraborty, D. & S.K. Das, 'Utilization and flow of energy in tropical ecosystem: A test case for two ox-bow lakes', *International Journal of Environmental Science and Technology*, 3 (2), 2006.

Chand, Dinesh, 'Towards Sustainable Rural Water Supply', *Kurukshetra*, October 2003.

Chand, Dinesh, 'Drinking Water Supply and Availability in rural areas' *Kurukshetra*, October 2002.

Chatterjee, N., *Irrigated Agriculture*, Rawat Publications, Jaipur/ Delhi 1995.

Choudhury, P. & P.K. Das, "Big Dams – Boon or curse in disguise – Tipaimukh and Subansiri experiences" in *PRAGYAN*

(ISSN No. 0976-4283) – A Quarterly Journal of Academic, Intellectual and Career pursuit, Published from Tinsukia College, Assam, Vol. VIII (No. 1), 2010.

Das, P.J., 'Rainfall regime of Northeast India : A hydrometeorological study with especial emphasis on the Brahmaputra basin', Ph.D thesis Gauhati University, Guwahati, 2004.

Das, P.J., 'Water and climate induced vulnerability in northeast India: Concerns for environmental security and sustainability', *WATCH Research Report 1, AARANYAK*, Guwahati, Assam, India. www.aaranyak.org 2009.

Das, P.J. & D.C. Goswami, 'Long Term variability of rainfall over northeast India', *Indian Journal of landscape systems and Ecological Studies* 26(1), 2003.

Das, T., 'A study on the limnology and capture fisheries of river Barak, Assam', M. Phil. Dissertation, Assam University, Silchar, 2001.

Das, T., A. Arunachalam, G. Gupta, & K. Arunachalam, 'Biodiversity of plankton and macroinvertebrates in Dikrong river system in the Indian Eastern Himalaya', In: A. Arunachalam & K. Arinachalam (eds.) *Biodiversity Utilization and Conservation*, Aavishkar Publishers & Distributors, Jaipur, 2007.

Das, T. & A. Gupta, 'An appraisal on the physico-chemical properties of water along a disturbance gradient in river Barak, Assam, north east India" *The Bioscan* 3(1), 2008.

Deka, T.K., M.M. Goswami & M. Kakati, 'Causes of Fish Depletion-A Factor Analysis Approach', *NAGA: World Fish Center Newsletter* 28 (1 & 2), 2005.

Devi, S.S., 'Ecology Of Baskandi Anua, An Oxbow Lake Of Cachar District, Assam, North East India', Unpublished M. Phil. dissertation, Assam University, Silchar, 2010.

Dey, M., 'Rain water Harvesting' in *Water Harvesting and Utilization in North East India*, Assam University Publication, 2007.

Drusilla, R., A. Kumaresan & M. Narayanan, 'Studies on water quality parameters of lotic systems in and around Courtallam (Tamil Nadu)', *Poll Res.*, Vol. 23, No. 3, 2004.

Duggal, V.K., 'Drinking Water and Sanitation', *Yojana*, June 2005.

Durgaprasad, P., 'Drinking Water; Management Strategies and Future Agenda', *Kurukshetra*, October 2003.

Duttagupta, S., S. Gupta & A. Gupta, 'Euglenoid blooms in the flood plain wetlands in Barak Valley, Assam, North Eastern India', *J. Environ. Biol.*, Vol. 25, No. 3, 2004.

Edmondson, W.T., *Fresh water Biology*, 2nd edn,, John Wiley and Sons, New York, 1959.

Foote L., S. Pandy & N.T. Krogman, 'Process of wetland loss in India', *Environmental Conservation*, 23, 1996.

Garg, J.K., T.S. Singh, & T.V.R. Murthy, *Wetlands of India*, Space Applications Centre (ISRO), Ahmedabad, India, 1998.

Garg, N.K. & Q. Hassan, 'Alarming scarcity of water in India', *Current Science* 93 (7), 2007.

Gleick, P. H., 'Water resources', In: *Encyclopedia of Climate and Weather*, ed. by S. H. Schneider, Oxford University Press, New York, vol. 2, 1996.

Goldman, C., 'Strategies for lake management in an increasing global environment', In Kumagai M. & Vincent, W.F. (ed.), *Freshwater management : global versus local perspectives*, Springer, Stuttgart, 2003.

Golterman, H.L., R.S. Clymo & M.A.M. Ohnstad, *Methods of Physical and Chemical Analysis of Freshwaters*, 2nd edn., Blackwell Scientific Publishers, Oxford, London, 1978.

Goswami, Dulal C., *Management of Water Resources of North-East India: Need for an Integrated Regional Plan* (www.indianfolklore.org), 2009.

Goswami, Pranay Jyoti, 'Status of Potable Water Supply in the Rural Areas of North East India with Special Reference to Barak Valley', in B. Datta Ray and R.P. Athparia (ed.),

Water and Water Resource Management, Omsons Publications, New Delhi, 2006.

Grizzetti, B., A. Lo Porto, L.J. Barkved, F. Bouraoui, J. Deelstra, & K.J. Joy, 'Modelling water pollution with stakeholders involvement – The twinned experience of Glomma (Norway) and Tungabhadra (India) river basins' Striver Policy Brief No. 10, 2008. [www.striver.no.]

Guha, U.C., *Kacharer Itivritta*, 1st edn. 1910, reprint Sadana Library , Dhaka, 1910.

Gupta, A., 'A comparative assessment of methods for biological surveillance of running water quality using macroinvertebrates', In: *Proc. 3rd National Symposium on Environment, Thiruvanthapuram*,(eds. N.B. Nair, C.D. Eapen, V.N. Bapat, S. Sadasivan, & P. Gangadharan,), Board of Research in Nuclear Sciences, Dept. of Atomic Energy, Govt of India, 1994.

Gupta, A., 'Streams of Shillong. An Ecological Overview', In: S. Sengupta, K.R. Sinha, S. Das, W.B. Rani & S. Purkayastha, *Rivers and Riverine Landscape in North East India*', Concept Publishing Company, New Delhi, 2006.

Gupta, A., "Drinking Water Quality in Barak Valley, Assam, North East India : Prioritizing Management Options', In *Proceedings of the 2nd APHW Conference*, Vol. I, Asia Pacific Association of Hydrology and Water Resources, Japan, 2004.

Gupta, A., 'Water Availability, Poverty and Socio-Economic Crisis in the Floodplains of Barak Valley, Assam, North East India', 2006. [https://www.mpl.ird.fr/ UR044/projets/ Textes/ Gupta. doc.].

Gupta, A., 'Water Harvesting as a Management Strategy in Agriculture and Public Health: The Barak Valley Scenario', in: N. Roy, D. Biswas & B.K. Dutta (eds.), *Water Harvesting and Water utilization in North East India*, pp. 108-118, Published by Assam University, 'Silchar, 2007.

Gupta, S., K. Buragohain & J.R. Bhuiyan, 'Importance of ponds in integrated water resource management in Barak Valley,

Assam, India', *Poll. Res.*, Vol. 27, No. 3, 2008.

Gupta, A. (2010), 'Forests and Wetlands of Barak Valley (Assam) : A Natural Capital Perspective', *Northeast Researches*, Journal of the Institute of Northeast India Studies, Vol. I, Kolkata, 2010.

Gupta, A. & R.G. Michael, (1983), 'Seasonal differences and relative abundance among population of benthic aquatic insects in a moderately high altitude stream'. *Proc. Wkshp. High. Alt. & Wildl. Ecol. Zool. Surv. India*, 1983.

Guy, D., *The ecology of the fish pond ecosystem with special reference to Africa*, Pergamon Press, 1992.

Hardin, G., 'The Tragedy of the Commons', *Science*, Vol. 162, Washington DC, 1978.

IFAD, DANIDA & UNOPS. *Report on Ox-bow Lake, Bangladesh*, Small Scale Fishermen Project: Mid Term Review, 1994.

IPCC, 'Summary for Policymakers', In: *Climate Change*, The Physical Science Basis. Contribution of Working Group I to the Fourth Assessment Report of the Intergovernmental Panel on Climate Change [S.D. Solomon, M. Manning Qin, Z. Chen, M. Marquis, K.B. Averyt, M.Tignor and H.L. Miller (eds.)]. Cambridge University Press, Cambridge, United Kingdom and New York, NY, USA, 2007.

ISI, *Draft Indian standard specification for drinking water*, Indian Standards Institution , DOC: CDC 26, 8626, New Delhi, 1982.

Irrigation Department, Assam, *The contribution made for creation of irrigationpotential*, Irrigation Department and Assam State Minor Irrigation Development Corporation, 2009.

IWMI, 'IWRM challenges in developing countries : lessons from India and elsewhere', *Water Policy Briefing – Issue 24*, International Water Management Institute, 2007. [http:// www. iwmi.cgiar.org/ Publications/ Water Policy Briefs/ PDF/WPB24.pdf].

Jønch-Clausen, T., *Integrated Water Resources Management (IWRM) and Water Efficiency by 2005: Why, What and*

How?, Global Water Partnership, Sweden, 2004.

IWMI, 'IWRM challenges in developing countries : lessons from India and elsewhere', *Water Policy Briefing – Issue 24*, International Water Management Institute, 2007. [http:// www. iwmi.cgiar.org/ Publications/ Water Policy Briefs/ PDF/WPB24.pdf].

Jønch-Clausen, T., *Integrated Water Resources Management (IWRM) and Water Efficiency by 2005: Why, What and How*?, Global Water Partnership, Sweden, 2004.

Kaushik S. & D.N. Saksena, 'Physio-Chemical Limnology of Certain Waterbodies of Central India', in: Vijoy Kumar K. (ed.) *Freshwater Eco System of India*, Daya Publishing House, Delhi, 1999.

Khan, T.I., 'Environmental Policies for Sustainable Development', *Environmental Policy Instruments*, Pointer Publishers, Jaipur, 2001.

Kumar, R., R.D. Singh, & K.D. Sharma, 'Water resources of India', *Current Science* 89 (5), 2005.

Laskar, H.S. & S. Gupta, 'Phytoplankton diversity and dynamics of Chatla floodplain lake, Barak Valley, Assam, North East India – A seasonal study' *J. Environ. Biol.*, Vol. 30, No.6, 2009.

Laskar, H.S. & S. Gupta, 'Ecology of a marsh in Chatla floodplain, Barak valley, North-East India', *Eco. Env. & Consv.* Vol. 16, No. 3, 2010.

Maddison, A., *Economic Progress and Policy in developing countries*, George Allen and Unwin Ltd., London, 1970.

Mahanta, Chandan, *Water Resources of the North East: State of the Knowledge Base*, Background paper No. 2, Indian Institute of Technology, Guwahati, India, March 2006. [http:/ /mdoner.gov.in/ writereaddata/ sublink3images/ Knowledge Base 3697746260.pdf].

Malmqvist, B. & S. Rundle, 'Threats to the running water ecosystems of the world', *Environ. Conserv.*, Vol. 29, No. 2, 2002.

Manasi, S., N. Latha, S. Paranjape, K.J. Joy, U.S. Nagothu, K.V. Raju, & P.P. Mollinga, (2009), 'Strategies and recommendations towards an IWRM approach in Tungabhadra sub-basin, India', Striver Policy Brief No. 1, 2005. [www.striver.no.].

Manohoran, S., V.K. Murugasan & R. Palaniswamy, 'Numerical Abundance of Benthic Macroinvertebrates in selected reservations of Tamil Nadu', *J. Inland Fish. Soc. India* 38, 2006.

McCartney, M, 'Living with dams: managing the environmental impacts', www.iwaponline.com.

Michael, P., *Ecological Methods for Field and Laboratory Investigations*, Tata McGraw-Hill Publishing Company Ltd., New Delhi, 1984.

Middendorp, H.A J., R. Hasan & N.A. Ahmed, 'Community fisheries management of Freshwater Lakes in Bangladesh', *NAGA*, The ICLARM Quarterly 19(1), 1996.

Mikulski, J., *Biology of Inland Waters*, PWN, Warszawa, 1974.

Mirza, M.M.Q., R.A. Warrick, N.J. Ericksen, & G.J. Kenny, 'Trends and persistence in precipitation in the Ganges, Brahmaputra and Meghna river basins', *Hydrological Sciences- Journal-des Hydrologiques*, Vol. 43(6), 1998.

Misra, Sib Ranjan, *Managing Canal Irrigation in India*, Concept Publishing House, New Delhi, 2006.

Moss B., I. Booker, H. Balls & K. Manson, 'Phytoplankton distribution in a temperate floodplain lake and river system. I. Hydrology, nutrient sources and phytoplankton biomass', *J. Plankton Res.*, Vol.11, 1919.

Mukhopadhyay, S.K., B. Chattopadhyay, A.R. Goswami & A. Chatterjee, 'Spatial variations in zooplankton diversity in waters contaminated with composite effluents', *J. Limnol.*, Vol. 66, No. 2, 2007.

Murthy, D.B.N., *Environmental Awareness and Protection*, Deep and Deep Publications Pvt. Ltd., New Delhi, 2004.

Nagothu, U.S. & S. Manasi, 'IWRM and livelihoods: fisheries in Tungabhadra sub-basin, India, Striver Policy Brief No. 1, 2008. [www.striver.no.].

National Water Policy, 2002, Ministry of Water Resources, Government of India, New Delhi, 2002.

National Habitation Survey, 2003.

Nautiyal, Suresh, 'Water Crisis Agenda for the Century', *Yojana*, February 2002.

Nwankwo, D.I. & S.A. Amuda, 'Periphyton Diatoms on three floating Aquatic Macrophytes in a polluted South-Western Nigerian Creek', *Intern. J.Ecol. Environ. Sci.*, Vol.19, 1993.

Offem, B.O., Y.A. Samsons, I.T. Omoniyi, & G.U. Ikpi, 'Dynamics of the limnological features and diversity of zooplankton populations of the Cross River System of SE Nigeria', *Knowl. Managt. Aquatic Ecosystem*, Vol. **393**, No. 02, 2009.

Palit, P. K., 'The History of Rivers in ancient Northern India', in N. Roy, B. K. Dutta & D.Biswas (ed.), *Water Harvesting and Water Utilization in North-east India*, Assam University Publication, Silchar, 2007.

Petrella, Recardo, 'Poverty, Water and Globilisation', *Yojana*, June 2005.

PHED, Govt. of Assam, *Gramyo Pani-jogan achonir porichalona swamitiupobidhi- 1995.*(Guidelines for managing committee of rural water supply scheme, 1995), Guwahati, 1995.

Pidwirny, M., *Fundamentals of Physical Geography*, 2nd Edition, 2006. [http://www. Physical geography. net/ fundamentals/ contents.html.].

Ponnuraj, S., 'Role of Panchayti Raj Institution in Drinking Water Supply Scheme', *Kurukshetra*, October 2003.

Rajora, R., 'Integrated Watershed Management', Philosophy and Concept of Watershed, Rawat Publications, Jaipur/New Delhi, 1998.

Rao, M.J. Mohan, 'Community Intervention in Traditional Rain Water Harvesting- Drinking Water', NIRD, Hyderabad, November 2003.

Rao, P.K., *Sustainable Development*, Blackwell Publishers, London, 2000.

Reddy, M.S. & N.V.V. Char, 'Management of Lakes in India: Theme paper in Lake management Initiative', India, 2004.

Renn, O., W.J. Burns., J.X. Kasperson, R.E. Kasperson & P. Slovic, 'The social amplification of risk: theoretical foundations and empirical applications', *Journal of Social Issues* 48(4), 1992.

Rosenberg, D.M. & V.H. Resh, *Freshwater biomonitoring and benthic invertebrates*, New York: Chapman and Hall, 1993.

Roy, N. & M.P. Bezbaruah, *Agricultural Growth and Regional Economic Development*, Mittal Publications, New Delhi, 2002.

Roy, Niranjan, 'Socio Economic Survey of Barak Valley (THEP), A Report of the Major Project', Funded by NEEPCO, Assam, 2009.

Roy, Niranjan, "Climate Change and Diversification of Agriculture in Northeast India", *Northeast Researches*, Journal of the Institute of Northeast India Studies, vol. I, Kolkata, 2010.

Saikia, K., 'Assessment of surface water quality and quantity in selected villages of Cachar and Karimganj district of Assam', Unpublished M Phil dissertation, Assam University, Silchar, 2009.

Sait, M. et. al., 'Positive and Negative Impacts of Dams on the Environment', www.dsi.gov.tr.

Salodia, P.K., 'Freshwater Biology An Ecological Approach, *Freshwater Ecology*, Surabhi Publications, Jaipur, 1996.

Sanghvi, Sheela, Pahwa Prem S., 'Water Harvesting, Purification and Distribution Management', *The Degradation of Freshwater Supplies*, pp.13-32, Dominant Publishers and Distributors, Daryaganj, New Delhi, 2001.

Saravanan, V.S., 'Integrated Water Resource Management: A Response', *Economic and Political Weekly*, September 23, 2006.

Sarma, H.P., 'Climate change and Water Resources of North East India', *Climate change, Global Warming and NE India, Regional perspective*, ERD Foundation, Guwahati, 2010.

Sethi, N., 'Tipaimukh Dam in Manipur driving a wedge?', *Down to Earth* (Oct, 15, 2006), Centre for Science and Environment, New Delhi, Vol. 16 (10), 2006.

Shah, T. & B. van Koppen, 'Is India ripe for integrated water resources management: fitting water policy to national development context', *Economic and Political Weekly* 41 (31), 2006.

Sharma, S.L.. 'Rethinking Sustainable Development – Ecology and Culture in India's Developmental Context, in: Choudhury, S.K. (ed.), *Culture, Ecology and Sustainable Development*, Mittal Publications, New Delhi, 2006.

Shaxson, T. F., 'People's Involvement in Watershed Management: Lessons from working among Resource-Poor Farmers' in: Lal R. (ed.), Integrated Watershed Management in the Global Ecosystem, pp. 345-361, CRC Press, LLC Florida, USA, 2000.

Singh, A.K., 'Arsenic contamination in the groundwater of North Eastern India', In: *Proceedings of National Seminar on Hydrology*, Roorkee, 2004.

Singh, K.P., 'Water Harvesting', *Yojana*, November 1996.

Sinha, M. & C. Jha, *Ecology and Fisheries of Ox-Bow Lakes (Maun) of North Bihar - A threatened ecosystem*: Bulletin 74, Central Inland Capture Fisheries Research Institute (ICAR), Barrackpore, West Bengal, 1997.

Sivakami, A., 'Uses of Water Resources – Ethics & Environment: A Case Study of Developing Countries', in: Kumar A. (ed.), *Ecology & Conservation of Lakes, Reserviors & Rivers*, ABD Publishers, Jaipur, 2002.

Snellen, W.B. & A. Schrevel, 'IWRM: for sustainable use of water – 50 years of international experience with the concept of integrated water management', *Background document to the FAO/Netherlands Conference on Water for Food and*

Ecosystems, Ministry of Agriculture, Nature and Food Quality, The Netherlands, 2004. [http:// www.fao.org/ ag/ wfe2005/ docs/ IWRM_Background.pdf].

Sparks, R. E., P.B. Bayley, S.L.Kohler & L.L. Osborne, 'Disturbance and recovery of large floodplain rivers', *Environmental Management* 14, 1990.

State of Environment Report Assam-2004, Assam State Technology and Environment Council.

Statistical Handbook, Govt. of Assam, Directorate of Economics and Statistics, Assam, Guwahati, 2005.

Swaminathan, D., 'Ensuring Supply of Safe Drinking Water', *Yojana*, January 26, 1995.

Thavaraj, S.H.C., 'The Importance of integrating non-engineering assets in irrigation system design' in Donald, C., Taylor & Thomas H. Wlckham (ed.), *Irrigation policy and the management of irrigation systems in south-east Asia*, The Agriculture Development Council Inc., Bangkok, 1982.

The Sangai Express, 7 February 2003; 'CM's dam statement flayed'.

The Imphal Free Press, 28 January 2003; 'Dam Opposition Opposed'.

The Sangai Express, 11 January 2003; 'AMUCO joins anti-dam chorus'.

'Tipaimukh: In the Quest for Survival', www. epao.com.

TOI, March 12, 2010, "33% water in India undrinkable".

United Nations Commission on Sustainable Development (UNCSD), *Comprehensive assessment of the Freshwater resources of the World*, UN, New York, 2000.

Uttah, E.C., C. Uttah, P.A. Akpan, E.M. Ikpeme, J. Ogbeche, L. Usip & J. Asor, 'Bio-survey of Plankton as indicators of water quality for recreational activities in Calabar River, Nigeria', *J. Appl. Sci. Environ. Managment.*, Vol.12, No. 2, 2008.

Verghese, B.G., 'Water Resources in the North East : Development options in a cooperative Framework'. WWW. North-East of India (PDF), 2006.

Verghese, B.G., 'Water resources in the Northeast: Development options in a cooperative framework', Background Paper No. 1, Draft for Discussion – March 2006, 2006. [http://mdoner.gov.in/ writereaddata/sublink3images/ cooperation 8437515003.pdf].

Water Resource Department, *A Report on Flood Damage in Assam*, Guwahati, 2004.

W.H.O., *Guideline for Drinking Water Quality*, 3rd Edition, World Health Organization, Geneva, 2004.

Winemiller, K.O. & K.A. Rose, 'Patterns of life history diversification in North American fishes: implications for population regulations', *Canadian Journal of Fisheries and Aquatic Sciences* 49, 1962.

World Commission on Environment and Development, *Our Common Future*, Oxford University Press, New York, 1987.

Yamamuro, M., I. Kioke & H. Iizuumi, 'Partitioning of nitrogen stocks in the vicinity of a Fijian seagrass bed dominated by *Syingodiun isoettifolum* (Ascherson) Dandy', *Australian Journal of Marine and Freshwater Resource*, Vol. 44, 1993.

Appendix–A

Participants of the Symposium

1. Professor J.B. Bhattacharjee, Chairman, Institute of Northeast India Studies, 13/A Udayan Abasan, Udayrajpur, Madhyamgram, Kolkata – 700129.
2. Professor Tapodhir Bhattacharjee, Vice-Chancellor, Assam University, Silchar -788011.
3. Professor G.D. Sharma, Pro-Vice Chancellor, Assam University, Silchar- 788011.
4. Professor Gautam Biswas, Pro-Vice Chancellor, Assam University, Silchar – 788011.
5. Professor Sib Ranjan Misra, Department of Economics, Visva-Bharati University, Santiniketan – 731235.
6. Professor B.P. Misra, Retd. Director, Centre for Himalayan Studies, North Bnegal University, Rammohanpur, Dist. Darjeeling, West Bengal, & Visiting Professor, Centre for Studies in Human Development, Assam University, Silchar – 788011.
7. Professor B.K. Dutta, Dean, School of Environmental Sciences, Assam University, Silchar – 788011.
8. Professor Abhik Gupta, Department of Ecology & Environmental Science, Assam University, Silchar – 788011.
9. Shri Rajdeep Chanda, Sr. Lecturer, Rabindra Sadan Girls' College, Karimganj – 788710.
10. Shri Nilanjan De, Lecturer, Rabindra Sadan Girls' College, Karimganj – 788710.

11. Dr. Pranay Jyoti Goswami, Associate Professor, Department of Commerce, Assam University, Silchar – 788011.
12. Ms. Rita Chakpram, Research Scholar, Department of Ecology & Environmental Science, Assam University, Silchar – 788011.
13. Shri Jibendu Dutta, D.I.E.T., Udharband, Dist. Cachar, Assam.
14. Shri Bimal Sinha, Research Scholar, Department of Business Administration, Assam University, Silchar – 788011.
15. Ms. Sharmila Naoseckpam, Research Scholar, Department of Ecology & Environmental Science, Assam University, Silchar – 788011.
16. Shri H. Khumarkima, Research Scholar, Department of Ecology & Environmental Science, Assam University, Silchar – 788011.
17. Dr. A. Das, Assistant Professor, Department of Bengali, Assam University, Silchar – 788011.
18. Dr. Hilloljyoti Singha, Assistant Professor, Centre for Biodiversity, Assam University, Silchar – 788011.
19. Shri Dilip Kumar Dey, Retd. Principal, T.T. College, Silchar – 788005.
20. Shri Jyotirmay Das, Retd. Chief Engineer, Agriculture, Govt. of Assam, Meherpur, Silchar – 788015.
21. Professor D.P. Roy, Senior Project Coordinator, CIDC Centre, Assam University, Silchar – 788011.
22. Shri Pijush Kanti Das, Research Scholar, Department of Bengali, Assam University, Silchar – 788011.
23. Dr. Parthankar Choudhury, Assistant Professor. Department of Ecology & Environmental Science, Assam University, Silchar – 788011.
24. Shri Asha Basumatary, Research Scholar, Department of History, Assam University, Silchar – 788011.
25. Shri Subhrajit Pyne, Department of Mathematics, Assam

University, Silchar – 788011.

26. Ms. Anupama Singha, Research Scholar, Department of History, Assam University, Silchar – 788001.
27. Ms. S. Pushpalata Singh, Research Scholar, Department of Economics, Assam University, Silchar – 788011.
28. Ms. Anuradha Singha, Research Scholar, Department of Economics, Assam University, Silchar – 788011.
29. Shri Sujay Das, Research Scholar, Department of Economics, Assam University, Silchar – 788011.
30. Shri Shamintra Ghosh, Research Scholar, Department of Economics, Assam University, Silchar – 788011.
31. Ms. Soma Dhar, Research Scholar, Department of Economics, Assam University, Silchar – 788011.
32. Shri Munim Ahmed Barbhuiya, Research Scholar, Department of Economics, Assam University, Silchar – 788011.
33. Shri Bhaskar Singha, Research Scholar, Department of Economics, Assam University, Silchar – 788011.
34. Ms. Kalpana Sen, Research Scholar, Department of History, Assam University, Silchar – 788011.
35. Dr. (Mrs) Mithra Dey, Reader, Department of Ecology & Environmental Science, Assam University, Silchar – 788011.
36. Dr. Mahmood Ansari, Assistant Professor, Department of Economics, Assam University, Silchar – 788011.
37. Dr. Niranjan Roy, Associate Professor, Department of Economic, & Director, IQAC, Assam University, Silchar-788011.
38. Dr. Panna Deb, Assistant Professor, Centre for Biodiversity & Natural Resource Conservation, Assam University, Silchar–788011.
39. Dr. (Mrs) Susmita Gupta, Assistant Professor, Department of Ecology & Environmental Science, Assam University, Silchar – 788011.

40. Dr. (Ms) Ratna Dey, Principal, D.E.I.T., Udharband, Dist. Cachar, Assam.

41. Ms. Hafsa Sultana Laskar, Research Scholar, Department of Ecology & Environmental Science, Assam University, Silchar – 788011.

42. Dr. (Ms.) Tapati Das, Assistant Professor, Department of Ecology & Environmental Science, Assam University, Silchar – 788011.

43. Dr. Manoj Kumar Sinha, Associate Professor & Head, Department of Library Science, Assam University, Silchar – 788001.

44. Dr. P. B. Mazumdar, Associate Professor, Department of Biotechnology, Assam University, Silchar – 788011.

45. Shri Sujay Sarkar, Department of History, Assam University, Silchar – 788011.

46. Ms. Babita Devi, Department of History, Assam University, Silchar – 788011.

47. Ms. Minky Choudhury, Department of History, Assam University, Silchar – 788011.

48. Shri Kanailal Bhattacharjee, Department of History, Assam University, Silchar – 788011.

49. Dr. K. K. Singh, Assistant Professor, Department of Ecology & Environmental Science, Assam University, Silchar – 788011.

50. Shri Shamikhu Chongmai, Department of Ecology & Environmental Science, Assam University, Silchar – 788011.

51. Shri Pankaj Chetri, Research Assistant, Bioinformatics Centre, Assam University, Silchar – 788011.

52. Dr. Avijit Debnath, Assistant Professor, Department of Economics, Assam University, Silchar – 788011.

53. Ms. Kankana Das, Research Scholar, Department of Economics, Assam University, Silchar – 788011.

54. Dr. Dipendu Das, Associate Professor & Head, Department

of English, Assam University, Silchar – 788011.

55. Dr. (Ms.) Sagarika Dey, Assistant Professor, Department of Economics, Assam University, Silchar – 788011.

56. Dr. Shabanti Maity, Assistant Professor, Department of Economics, Assam University, Silchar – 788011.

57. Professor Nikhil Bhusan Dey, Department of Commerce, Assam University, Silchar – 788001.

58. Professor D.K. Pandiya, Department of Commerce, Assam University, Silchar – 788011.

59. Professor G.P. Pandey, Department of Mass Communication, Assam University, Silchar – 788011.

60. Dr. Ramakanta Das, Assistant Professor, Department of Bengali, Assam University, Silchar – 788011.

61. Dr. Samira Behera, Assistant Professor, Department of Mathematics, Assam University, Silchar – 788011.

62. Dr. (Ms.) Sudeshna Purkayastha, Associate Professor & Head, Department of History, Assam University, Silchar – 788011.

63. Dr. Projit Kumar Palit, Associate Professor, Department of History, Assam University, Silchar – 788011.

64. Dr. (Mrs) Suparna Roy, Assistant Professor, Department of History, Assam University, Silchar – 788011.

65. Dr. M. Dutta Choudhury, Associate Professor, Department of Life Science, Assam University, Silchar – 788011.

66. Dr. S.B. Paul, Associate Professor, Department of Chemistry, Assam University, Silchar – 788011.

67. Dr. B.R. Choudhury, Director, College Development Council, Assam University, Silchar – 788001.

68. Dr. Bibhas Deb, Deputy Registrar, Assam University, Silchar – 788001.

69. Shri Ananta Pegu, Research Scholar, Department of Economics, Assam University, Silchar – 788011.

70. Ms. Kankana Das, Research Scholar, Department of Ecology & Environmental Science, Assam University, Silchar – 788011.

Index

□□□